Springer Theses

Recognizing Outstanding Ph.D. Research

For further volumes:
http://www.springer.com/series/8790

Aims and Scope

The series "Springer Theses" brings together a selection of the very best Ph.D. theses from around the world and across the physical sciences. Nominated and endorsed by two recognized specialists, each published volume has been selected for its scientific excellence and the high impact of its contents for the pertinent field of research. For greater accessibility to non-specialists, the published versions include an extended introduction, as well as a foreword by the student's supervisor explaining the special relevance of the work for the field. As a whole, the series will provide a valuable resource both for newcomers to the research fields described, and for other scientists seeking detailed background information on special questions. Finally, it provides an accredited documentation of the valuable contributions made by today's younger generation of scientists.

Theses are accepted into the series by invited nomination only and must fulfill all of the following criteria

- They must be written in good English.
- The topic should fall within the confines of Chemistry, Physics, Earth Sciences, Engineering and related interdisciplinary fields such as Materials, Nanoscience, Chemical Engineering, Complex Systems and Biophysics.
- The work reported in the thesis must represent a significant scientific advance.
- If the thesis includes previously published material, permission to reproduce this must be gained from the respective copyright holder.
- They must have been examined and passed during the 12 months prior to nomination.
- Each thesis should include a foreword by the supervisor outlining the significance of its content.
- The theses should have a clearly defined structure including an introduction accessible to scientists not expert in that particular field.

Kosuke Nomura

Interacting Boson Model from Energy Density Functionals

Doctoral Thesis accepted by
the University of Tokyo, Tokyo, Japan

Author (Current address)
Dr. Kosuke Nomura
Institut für Kernphysik
Universität zu Köln
Köln
Germany

Supervisor
Prof. Takaharu Otsuka
Department of Physics
Graduate School of Science
The University of Tokyo
Tokyo
Japan

ISSN 2190-5053
ISBN 978-4-431-54233-9
DOI 10.1007/978-4-431-54234-6
Springer Tokyo Heidelberg New York Dordrecht London

ISSN 2190-5061 (electronic)
ISBN 978-4-431-54234-6 (eBook)

Library of Congress Control Number: 2012954856

© Springer Japan 2013

This work is subject to copyright. All rights are reserved by the Publisher, whether the whole or part of the material is concerned, specifically the rights of translation, reprinting, reuse of illustrations, recitation, broadcasting, reproduction on microfilms or in any other physical way, and transmission or information storage and retrieval, electronic adaptation, computer software, or by similar or dissimilar methodology now known or hereafter developed. Exempted from this legal reservation are brief excerpts in connection with reviews or scholarly analysis or material supplied specifically for the purpose of being entered and executed on a computer system, for exclusive use by the purchaser of the work. Duplication of this publication or parts thereof is permitted only under the provisions of the Copyright Law of the Publisher's location, in its current version, and permission for use must always be obtained from Springer. Permissions for use may be obtained through RightsLink at the Copyright Clearance Center. Violations are liable to prosecution under the respective Copyright Law.

The use of general descriptive names, registered names, trademarks, service marks, etc. in this publication does not imply, even in the absence of a specific statement, that such names are exempt from the relevant protective laws and regulations and therefore free for general use.

While the advice and information in this book are believed to be true and accurate at the date of publication, neither the authors nor the editors nor the publisher can accept any legal responsibility for any errors or omissions that may be made. The publisher makes no warranty, express or implied, with respect to the material contained herein.

Printed on acid-free paper

Springer is part of Springer Science+Business Media (www.springer.com)

Publications

- K. Nomura, N. Shimizu, and T. Otsuka, *Mean-field Derivation of the Interacting Boson Model Hamiltonian and Exotic Nuclei*, Phys. Rev. Lett. **101**, 142501 (2008).
- K. Nomura, N. Shimizu, and T. Otsuka, *Formulating the interacting boson model by mean-field methods*, Phys. Rev. C **81**, 044307 (2010).
- K. Nomura, T. Otsuka, N. Shimizu, and L. Guo, *Microscopic formulation of the interacting boson model for rotational nuclei*, Phys. Rev. C **83**, 041302(R) (2011).
- K. Nomura, T. Otsuka, R. Rodríguez-Guzmán, L. M. Robledo, and P. Sarriguren, *Structural evolution in Pt isotopes with the interacting boson model Hamiltonian derived from the Gogny energy density functional*, Phys. Rev. C **83**, 014309 (2011).
- K. Nomura, T. Otsuka, R. Rodríguez-Guzmán, L. M. Robledo, and P. Sarriguren, P. H. Regan, P. D. Stevenson, and Zs. Podolyák, *Spectroscopic calculations of the low-lying structure in exotic Os and W isotopes*, Phys. Rev. C **83**, 051303 (2011).
- K. Nomura, *Microscopic derivation of IBM and structural evolution in nuclei*, AIP Conference Proceedings **1355**, "International Symposium; New Faces of Atomic Nuclei", 23–28 (2011).
- K. Nomura, T. Nikšić, T. Otsuka, N. Shimizu, and D. Vretenar, *Quadrupole collective dynamics from energy density functionals: Collective Hamiltonian and the interacting boson model*, Phys. Rev. C **84**, 014302 (2011).
- K. Nomura, T. Otsuka, R. Rodríguez-Guzmán, L. M. Robledo, and P. Sarriguren, *Collective structural evolution in Yb, Hf, W, Os and Pt isotopes*, Phys. Rev. C **84**, 054316 (2011).
- M. Albers, N. Warr, K. Nomura, A. Blazhev, J. Jolie, D. Mücher, B. Bastin, C. Bauer, C. Bernards, L. Bettermann, V. Bildstein, J. Butterworth, M. Cappellazzo, J. Cederkäll, D. Cline, I. Darby, S. Das Gupta, J. M. Daugas, T. Davinson, H. De Witte, J. Diriken, D. Filipescu, E. Fiori, C. Fransen, L. P. Gaffney, G. Georgiev, R. Gernhäuser, M. Hackstein, S. Heinze, H. Hess, M. Huyse, D. Jenkins, J. Konki, M. Kowalczyk, T. Kröll, R. Lutter, N. Marginean, C. Mihai, K. Moschner, P. Napiorkowski, B. S. Nara Singh, K. Nowak, T. Otsuka, J. Pakarinen, M. Pfeiffer, D. Radeck, P. Reiter, S. Rigby, L. M. Robledo, R. Rodríguez-Guzmán, M. Rudigier, P. Sarriguren, M. Scheck, M. Seidlitz, B. Siebeck, G. Simpson, P. Thoele, T. Thomas, J. Van de Walle, P. Van Duppen, M. Vermeulen, D. Voulot, R. Wadsworth, F. Wenander, K. Wimmer, K. O. Zell, and M. Zielinska, *Evidence for a smooth onset of deformation in the neutron-rich Kr isotopes*, Phys. Rev. Lett. **108**, 062701 (2012).
- K. Nomura, N. Shimizu, D. Vretenar, T. Nikšić, and T. Otsuka, *Robust regularity in γ-soft nuclei and its novel microscopic realization*, Phys. Rev. Lett. **108**, 132501 (2012).

Supervisor's Foreword

The interacting boson model, proposed by Prof. Akito Arima and Prof. Francesco Iachello in 1975, is one of the major models of nuclear physics. It has been applied very successfully to the description of low-energy nuclear quadrupole collective states. These quantum states possess characteristic features of the nuclear collective motion, such as surface vibration, rotation of rigid ellipsoids, and their intermediate situations including triaxial deformation. Thus, the collective motion is one of the primary subjects of nuclear physics with fundamental importance. The model exploits its algebraic structure, and various aspects of the collective states have been explored.

Despite such success, what has been missing is the microscopic justification or foundation for all situations of the collective states. Here, by "all situations" I mean vibrational, rotational, and their intermediate situations mentioned above. The microscopic foundation implies an explanation or derivation of the model from an underlying microscopic system that is nothing but nucleon systems here. This implies the correspondence between collective pairs of valence nucleons and bosons in the interacting boson model. It was created for nearly spherical and weakly deformed cases, including those in-between, by myself in collaboration with Arima and Iachello in 1978. Although this work was one of the major steps in the development of the interacting boson model, a unified foundation of the model covering all cases was still to come.

The work presented in this thesis has filled such a gap and has paved the way toward a unified description of the nuclear low-energy collective motion by providing us with a beautiful, strong bridge from nucleon systems to boson systems of the interacting boson model. On the other hand, the validity of this work is based on that of the microscopic energy density functional framework. Indeed, the self-consistent mean-field theory with a given energy density functional currently provides an accurate and global description of nuclear bulk properties and collective excitation over almost the whole range of the nuclear chart, and is also quite suitable to start with to construct the bridge. I would like to point out here that the interacting boson model and the energy density functional framework can and should develop collaboratively and complementarily.

This thesis shows a unified framework of the microscopic basis of the model and exhibits various features arising from this new framework—for instance, the triaxial shape and its description by three-boson interaction. The actual outcome, conceptual and numerical, is shown in detail in the text, and I will not mention it here. I would like to restrict myself to mentioning that the work presented here has solved problems and debates that remained open for more than 30 years, and that Dr. Nomura has made the solutions possible in 5 years as a Ph.D. student. The work presented in this thesis is a truly amazing achievement and deserves this special publication.

Tokyo, August 2012 Takaharu Otsuka

Acknowledgments

The author is heartily grateful to Prof. T. Otsuka and Prof. N. Shimizu, whose encouragement, guidance, and support from the initial to the final levels enabled him to develop an understanding of the subject. The author would like to thank Prof. Otsuka for the generous financial supports for numerical work, attending workshops, and visiting other institutes abroad, without which this work could not be completed as it is. The author thanks Dr. M. Albers, Dr. L. Guo, Dr. T. Nikšić, Dr. Zs. Podolyák, Prof. P. H. Regan, Prof. L. M. Robledo, Dr. R. Rodríguez-Guzmán, Prof. P. Sarriguren, Dr. P. D. Stevenson, and Prof. D. Vretenar for stimulating discussions and collaborations, thanks Prof. A. Arima, Prof. P. von Brentano, Prof. R. F. Casten, Prof. A. Gelberg, Prof. F. Iachello, Prof. J. Jolie, Dr. D. Lacroix, Prof. T. Mizusaki and Dr. P. Van Isacker for valuable comments on this work and for fruitful discussions, and is also grateful to his colleagues in the nuclear theory group of the University of Tokyo for continuous encouragement and help. Finally the author acknowledges the support by the Japan Society for the Promotion of Science during the last three years of his Ph.D. course.

Contents

1 Introduction 1
References 8

2 Basic Notions 15
2.1 General Remarks 15
2.2 Self-Consistent Mean-Field Models 16
    2.2.1 Density-Dependent Force 17
    2.2.2 Constrained Mean Field 22
2.3 Interacting Boson Model 24
    2.3.1 Algebras 24
    2.3.2 Geometry 27
    2.3.3 Hamiltonian 29
    2.3.4 Other Boson Models 31
2.4 Nucleon-to-Boson Mapping 32
    2.4.1 Optimal Boson Hamiltonian 33
    2.4.2 Uniqueness of the Boson Parameters 43
2.5 Brief Summary 48
References 49

3 Rotating Deformed Systems with Axial Symmetry 53
3.1 A Piece of History, and Basics 53
3.2 Rotational Cranking 54
3.3 Results and Discussions 59
    3.3.1 Rotational Bands 59
    3.3.2 Validity of Cranking Formula 61
3.4 Brief Summary 62
References 63

4	**Weakly Deformed Systems with Triaxial Dynamics**		65
	4.1 Quantum Phase Transitions		65
	4.2 Axially to γ-Unstable Deformed Nuclei		68
		4.2.1 $Z < 50$, $50 \leq N \leq 82$ Major Shells	69
		4.2.2 $Z > 50$, $50 \leq N \leq 82$ Major Shells	74
		4.2.3 E(5) Symmetry in Exotic Nuclei	80
	4.3 Prolate-Oblate Shape Dynamics		84
		4.3.1 IBM from Gogny D1S	84
		4.3.2 Evidence for Critical Points	97
		4.3.3 Systematics from Gogny-D1M Functional	105
		4.3.4 QPT in Exotic Nuclei	119
	References		124
5	**Comparison with Geometrical Model**		131
	5.1 Aim		131
	5.2 Bohr Hamiltonian		132
	5.3 Geometrical and Bosonic Spectra		137
	5.4 Brief Summary		140
	References		142
6	**Is Axially Asymmetric Nucleus γ Rigid or Unstable?**		145
	6.1 Overview		145
	6.2 Three-Body Boson Term		147
	6.3 Brief Summary		153
	References		154
7	**Ground-State Correlation**		157
	7.1 Binding and Two-Neutron Separation Energies		157
	7.2 Empirical Proton-Neutron Correlation		161
	7.3 Brief Summary		162
	References		164
8	**Summary and Concluding Remarks**		165
	References		169

Appendix A: Details of Mean-Field Calculations ... 171

Appendix B: Formulas in the IBM-2 Framework ... 177

Abbreviations

BCS	Bardeen-Cooper-Schrieffer
CERN	Conseil Européen pour la Recherche Nucléaire
DFT	Density Functional Theory
EDF	Energy Density Functional
FRIB	Facility for Rare Isotope Beams
GANIL	Grand Accélérateur National d'Ions Lourds
GCM	Generator Coordinate Method
GSI	Gesellschaft für Schwerionenforschung
HF	Hartree-Fock
HFB	Hartree-Fock-Bogoliubov
IB	Inglis-Belyaev
IBM	Interacting Boson Model
IBM-1	Interacting Boson Model-1
IBM-2	Interacting Boson Model-2
MF	Mean Field
MOI	Moment of Inertia
OAI	Otsuka-Arima-Iachello
PES	Potential Energy Surface
RHB	Relativistic Hartree Bogoliubov
RIKEN	The Institute of Physical and Chemical Research
RMF	Relativistic Mean-field
TRIUMF	Canada's National Laboratory for Particle and Nuclear Physics
TV	Thouless-Valatin
QPT	Quantum Phase Transition

Chapter 1
Introduction

Atomic nucleus is a highly quantal-mechanical, finite many-body system comprised of protons and neutrons, where the strong, the weak and the electromagnetic fundamental interactions play an important role at the most profound level. The study of the atomic nucleus has been therefore crucial for elucidating the origin of matter (or nucleosynthesis processes), the tests of fundamental symmetries, and even the purpose of practical applications. Furthermore the way to understand the structure of nucleus is interdisciplinary since it applies to other fields of mesoscopic quantum systems such as condensed matter, atomic and polyatomic molecular physics. Thanks to the rigorous experimental efforts worldwide that make use of a new generation of rare-isotope beams e.g., at RIKEN in Japan, FRIB and TRIUMF in North America, CERN, GANIL and GSI in Europe etc, it has nowadays become possible to produce and to accelerate extremely unstable, i.e., short-lived, nuclei with considerable proton or neutron excess. The nuclei under such extreme conditions present many unexpected facets, and are therefore called *exotic nuclei*.

Since the pioneering work by Mayer and Jensen [1, 2], formation of shell structure has been one of the remarkable features of atomic nucleus in understanding the nuclear structure. In what is called independent-particle (or shell) model, a nucleon in the nucleus is taken as being moving with an average potential created by all other nucleons. This is much alike the dynamics of electrons in an atom, and similarly to these exhibits discrete single-particle energies. When protons and/or neutrons are filled from the lowest- up to the higher-lying orbitals to reach specific values like 2, 8, 20, 28, 50, 82, 126, ..., then a nucleus is notably stable and hence large amount of energy is needed to excite the nucleus from the closed shell to the next. These numbers are called magic numbers, which become evident as a sudden drop of the observed nucleon separation energies. In exotic nuclei, conventional magic numbers may become no longer valid, even giving rise to novel shell structures not heretofore recognized.

Besides these intriguing features that reflect single-nucleon degrees of freedom, the nucleus as a whole exhibits collective properties associated with a distinct shape, where all the constituent nucleons are coherently involved. The collective motion,

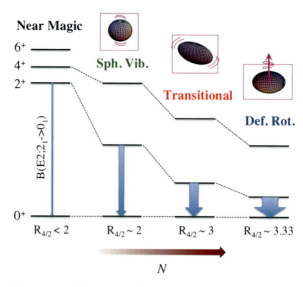

Fig. 1.1 Pictorial description of the quadrupole collective states of atomic nucleus. When departing from the closed shell (Near Magic) with the increase of the valence nucleon number N, the shape changes from spherical vibrator (Sph. Vib.) to deformed rotor (Def. Rot.), passing through the transitional nuclei in between. Each shape results in the characteristic level structure: phonon-like level scheme for a vibrator, and a clear rotational band for a rotor, which are well indicated by the ratio of 4_1^+ to 2_1^+ excitation energies, denoted by $R_{4/2}$. As the collectivity evolves with the number of valence nucleons, the electric quadrupole (E2) transition intensity from the 2_1^+ excited state to the 0_1^+ ground state becomes stronger

observed normally in low-energy[1] regime, stems from the deformation of nuclear surface, that is induced by the multi-fermion dynamics [3–7]. The microscopic interpretation on such a nuclear collective motion was already come up with by Rainwater in 1950 [3], on top of which Bohr and Mottelson established an well-known geometrical model in the middle of 1950s [4–7]. The collective model incorporates the single-particle (shell-model) feature into the purely classical description of the intrinsic nuclear shape as a macroscopic droplet. Particularly the most basic, yet significant nuclear collective motion can be of quadrupole type: the shape of a nucleus can be a spherical vibrator, an ellipsoidal deformed rotor and an object in between, depending on the number of active nucleons. Consequently, a class of remarkable regularities emerge in the corresponding spectroscopic properties (cf. Fig. 1.1). Deformation occurs as a consequence of the intrinsic spontaneous symmetry breaking of the nuclear mean field (in analogy with Jahn-Teller effect [8]), and the rotational motion manifests itself as a realization of symmetry-restoration mechanism [9, 10], which is highly relevant to understanding the microscopy of the nuclear quadrupole deformation.

---

[1] The energy scale for the collective mode of excitation is typically of the order of 1–10 MeV.

1 Introduction

The nucleus is a strongly correlating system governed by the complex nuclear force acting among individual nucleons, while it is, as a whole, a self-bound object characterized by a rather distinct shape seen through the regular patterns of the collective excitations. Therefore, to understand the regularities of the collective mode of excitation from a more microscopic degree of freedom has been a theme of major interest in nuclear physics [3–7, 11–20]. The purpose of this thesis is to address this issue from the viewpoint of the interacting boson model [16, 17] that is formulated by microscopic nuclear energy density functionals. Note that *microscopic* in this context refers to the single-nucleon degrees of freedom, and that we basically assume nucleons (both protons and neutrons) as elementary degrees of freedom throughout this thesis.

Microscopic studies based on nuclear energy density functionals (EDFs) have been quite successful in reproducing with remarkable accuracy various intrinsic (bulk) properties of almost all medium-mass and heavy nuclei on the periodic table such as binding energies, density distributions, surface deformations, charge radii, giant resonances, etc [14, 15]. The current and well-established generation of EDFs includes non-relativistic Skyrme- [21–24], which is of zero-range nature, and Gogny- [25, 26], which is of finite-range type, functionals as well as other density functionals associated with the relativistic mean-field Lagrangian of the effective theory of two-flavor quantum chromodynamics [27–29]. The framework of EDFs has also been extended beyond the mean-field level to describe excitation spectra and electromagnetic transition rates. Models have been developed that perform restoration of symmetries broken by the static nuclear mean field, and take into account quadrupole fluctuations: configuration mixing calculations in the spirit of the generator coordinate method [14, 15, 30–38], and solutions of the Bohr-type collective model Hamiltonian with quadrupole degrees of freedom [39–43].

A static self-consistent mean-field solution in the intrinsic frame, for instance a map of the energy surface as a function of quadrupole deformation, is characterized by symmetry breaking: translational, rotational, particle number, and can only provide an approximate description of bulk ground-state properties. To calculate excitation spectra and electromagnetic transition rates in individual nuclei, it is necessary to include correlations that arise from symmetry restoration and fluctuations around the mean-field minimum. Both types of correlations can be included simultaneously by mixing angular-momentum projected states corresponding to different quadrupole moments. The most effective approach for configuration mixing calculations is the generator coordinate method (GCM), with multipole moments used as coordinates that generate the intrinsic wave functions. It must be noted that, while GCM configuration mixing of axially symmetric states has been implemented by several groups and routinely used in nuclear structure studies [44–47], the application of this method to triaxial shapes presents a much more involved and technically difficult problem [33, 38]. In addition, the use of general EDFs, that is, with an arbitrary dependence on nucleon densities, in GCM type calculations, often leads to discontinuities or even divergences of the energy kernels as a function of deformation [48, 49]. Only for certain types of density dependence a regularization method can be implemented, which corrects energy kernels and removes the discontinuities and divergences [50–52].

As a sound approximation to the full GCM approach to five-dimensional quadrupole dynamics that restores rotational symmetry and that allows for fluctuations around the triaxial mean-field minima, a collective Hamiltonian can be formulated, with deformation-dependent parameters determined by constrained microscopic self-consistent mean-field calculations. The dynamics of the five-dimensional Hamiltonian for quadrupole vibrational and rotational degrees of freedom is governed by the seven functions of the intrinsic quadrupole deformations: the collective potential, three vibrational mass parameters, and three moments of inertia for rotations around the principal axes [39–43].

Another successful approach to the low-lying structure of medium-heavy and heavy nuclei consists in mapping[2] of the multi-nucleon dynamics onto the appropriate system of interacting bosons [16, 17]. The interacting boson model (IBM) of atomic nucleus, originally invented by Arima and Iachello [16, 17], has witnessed great deal of success for the phenomenological description of the low-lying quadrupole collective states of medium-heavy and heavy nuclei. The main ansatz of IBM is to employ the $J^\pi = 0^+$ (s) and 2^+ (d) bosons, which are supposed to simulate the motion of the collective nucleon pairs coupled to angular momentum $J^\pi = 0^+$ and 2^+, respectively, and to introduce the relevant interactions between the bosons [53, 54]. The IBM embodies an entire class of symmetries and regularities of the low-lying quadrupole collective states: three dynamical symmetries arising from the spontaneous breaking of U(6) symmetry, U(5) [55], SU(3) [56] and O(6) [57] limits, where the boson Hamiltonian can be written in some specific forms based on simple algebraic relations, and the intermediate situations of these limits, to which most realistic nuclei belong. The IBM, as well as its algebraic feature, is so general that it has been applied not only in but outside of nuclear physics [16, 17, 58, 59], and thus is itself of wide interest. The IBM in its earliest version (referred to as IBM-1) is purely phenomenological so that the interaction strengths of the model Hamiltonian have been determined from experiment or taken from earlier fitting calculations. Therefore, the IBM itself should have a certain microscopic foundation starting from the nucleonic degrees of freedom.

From a microscopic viewpoint [53, 54, 60], the IBM is essentially a vast truncation of the nuclear shell model, where the so-called proton monopole s_π and quadrupole d_π bosons and neutron monopole s_ν and quadrupole d_ν bosons reflect collective pairs of valence protons, S_π and D_π, and neutrons, S_ν and D_ν, respectively. As the numbers of valence protons and neutrons are constant for a given nucleus, the numbers of proton and neutron bosons, denoted respectively by N_π and N_ν, are set equal to half of the valence proton and neutron numbers. The interaction strengths of the boson Hamiltonian have been determined by the mapping from the SD subspace of the full shell-model space onto the sd boson space. The mapping scheme for deriving the IBM Hamiltonian of this type is usually referred to as the Otsuka-Arima-Iachello (OAI) mapping and can be extended as the proton-neutron interacting boson model (IBM-2) as a natural consequence [53, 54]. The OAI mapping has been practiced for limited realistic cases of nearly spherical or γ-unstable

---

[2] Further explanation of the terminology "mapping" will be given in Sect. 2.4.1.

1 Introduction 5

shapes [61–64] by using zero- and low-seniority states of the shell model [53, 54, 60], and has been also tested for deformed Sm isotopes by renormalizing the contribution from the G-pairs as a perturbation [65]. A fermion-boson mapping for deformed nuclei has been studied partly by the "independent-pair" property of condensed coherent fermion pairs [66] and by the rotation of the intrinsic state (a state in the body-fixed frame) [67]. In addition, there are many systematic calculations within the IBM-2 phenomenology for, e.g., Xe-Ba-Ce [68], Ru-Pd [69], Kr [70] and W-Os [71, 72] regions. The microscopic basis of the IBM has been studied for many years, but is still an open problem for the cases involving the strongly deformed nuclei.

More recently a general way of deriving the Hamiltonian of IBM-2 was proposed by Nomura et al. [73]. Under the assumption that the multi-fermion dynamics of the surface deformation is simulated by effective bosonic degrees of freedom, the energy expectation value with varying quadrupole deformation (so-called potential energy surface; PES) within the self-consistent mean-field calculation with a fixed microscopic EDF is mapped onto the corresponding classical limit of the appropriate boson Hamiltonian. Energies and wave functions of excited states are yielded with good angular momentum and particle number [73, 75]. As a given EDF allows universal description of the nuclear intrinsic properties including deformation of ground-state shape, this mapping process in principle provides the interaction strengths of the IBM Hamiltonian for any situations of the quadrupole collective states. While any popular EDF has a direct correspondence to the quadrupole deformation and is certainly suitable to start with, the IBM is a model for nuclear spectroscopy, that provides almost complete description of low-lying structure in medium-heavy and heavy nuclei and that embodies relevant physics in a straightforward way. Therefore we try to incorporate a successful EDF approach in the IBM framework. The validity of the initial work of Ref. [73] was further examined in Ref. [75]: the uniqueness of the derived parameters have been examined carefully using the method of the Wavelet transform [76].

When it is formulated microscopically, however, the IBM is shown to have a crucial problem of not capable of reproducing the moment of inertia of rotational band of strongly deformed nuclei. The problem occurs also in the new scheme of Ref. [73]: the moment of inertia calculated by the IBM turns out be by several tens per cent smaller than the experimentally observed one. Originally, the issue arose as a consequence of the critical comment made by Bohr and Mottelson in 1980, based on a microscopic theory using Nilsson plus BCS model [77]. They concluded that the SD truncation might not be sufficient to account for the intrinsic state of rotational deformed nuclei. This question should lead to the problem concerning whether or not the sd-IBM can be justified for deformed nuclei. In spite of considerable amounts of theoretical works for the past decades concerning the critique by Bohr and Mottelson, any conclusive work that justifies the validity of IBM for rotational motion has been missing. An important piece of information as to the critique was provided recently by Nomura et al. [78]. They suggested that the deformed nucleon system is substantially different in its response to infinitesimal rotation (cranking) from the corresponding deformed boson system.[3] It was shown [78] that, when the difference in the rotational response becomes sizable, then it can be a possible microscopic origin of the problem concerning the rotational moment of inertia. To correct the difference in the rotational response between fermion and boson systems, the rotational kinetic-like term (so-called LL term) was introduced in the boson system. As a consequence, the rotational bands of strongly-deformed rare-earth and actinoid nuclei were reproduced almost perfectly without any phenomenological adjustment. This study revisited the criticism made in the past by Bohr and Mottelson, and showed, for the first time, how the IBM can be justified for rotational motion of strongly deformed nuclei.

In most isotopic or isotonic sequences the transition between different shapes is gradual, but in a number of cases, with the addition or subtraction of only few nucleons, one finds signatures of abrupt changes in observables that characterize equilibrium shapes. These structure phenomena have been investigated using concepts of quantum shape/phase transitions in finite nuclear system [18, 19, 79–82], and advanced self-consistent (beyond) mean-field approaches [15, 37, 42, 43, 83–94]. In particular, the complex interplay between several deformation degrees of freedom, taking place in different regions of the nuclear chart, offers the possibility of testing microscopic descriptions of atomic nuclei under a wide variety of conditions. In this context, mean-field approximations based on effective EDFs, which as shown already are a cornerstone to almost all microscopic approximations to the nuclear many-body problem, appear to be a first tool to rely on when looking for fingerprints of nuclear shape/phase transitions. On the other hand, it has also become possible to recast mean-field equations in terms of efficient minimization procedures such as the so-called gradient method [95, 96]. One of the advantages of the gradient method is the way it handles constraints, which is well adapted to the case where a large number of constraints are required (like the case which requires, in addition to the proton and neutron number constraints, constrains on both β and γ degrees of freedom characterizing the nuclear shape). Another advantage is its robustness in reaching a solution, a convenient property when large scale calculations requiring the solution of many HFB equations are performed. Experimentally, low-lying spectroscopy provides one with a very powerful source of information that allows establishing signatures correlating nuclear shape transitions with excitation spectra [97–107]. Along these works, the method of [73] has been already tested in a number of spectroscopic calculations in order to clarify the collective structural evolution in various mass regions: Neutron-rich Kr isotopes with mass $A \approx 90$–100 [108], Ru-Pd isotopes with $A \approx 100$–120 [75], Ba-Xe isotopes with $A \approx 110$–130 [75], Sm-Gd isotopes with $A \approx 150$ [73, 75, 78], Pt [109] and Os-W [110] isotopes with $A \approx 180$–200, as well as more systematic analysis on Yb-Hf isotopes in addition to the last three in the same mass region [111].

What is also of interest concerns whether the IBM Hamiltonian, derived from an EDF, can have equal predictive power as other EDF-based schemes, such as

---

[3] The rotational response in this context means the change of the ground-state energy due to the infinitesimal rotation.

the collective Hamiltonian approach. In Ref. [112], the spectroscopic observables resulting from the IBM-2 Hamiltonian were compared with the solutions of the five-dimensional collective Hamiltonian, with both models starting from the density-dependent point-coupling interaction (DD-PC1) [113] of the relativistic Hartree-Bogoliubov model. The comparison of the two schemes has been done in heavy Pt isotopes, and it was shown in Ref. [112] that both methods do work similarly quite well in the ground-state band spectra but that a certain difference between the two prescriptions comes out e.g., in the structure of the quasi-γ band and in the E2 transition pattern within the ground-state band.

Meanwhile, the structure of non-axial nuclei has been described by the two major geometrical models: the rigid-triaxial rotor model of Davydov and Filippov [114] and the γ-unstable rotor model of Wilets and Jean [115]. However, presumably all observed non-axial medium-heavy and heavy nuclei fall exactly in between the rigid-triaxial and the γ-unstable rotor pictures. This puzzle was addressed in Ref. [116], which showed that, based on a microscopic energy density functional calculation, neither of the rigid-triaxial nor γ-unstable rotor descriptions is realized in actual nuclei. This empirically known fact can be explained naturally only with the inclusion of the three-body boson term into the IBM-2 system, and is shown to be independent of the choice and the details of the EDFs. The result also points to the most appropriate IBM description of γ-soft systems.

This thesis is organized as follows: Chap. 2 explains the *proof of principle*, i.e., the way to determine the IBM Hamiltonian by the EDF approach, as well as its physical interpretations. Crucial limitation inherent to the microscopic IBM, which one encounters in reproducing the moment of inertia of rotational band, is pointed out. This naturally casts a question as to the validity of IBM for deformed nuclei, and a possible answer to this question is proposed in Chap. 3. In Chap. 4, spectroscopic calculations are presented for sets of medium-heavy and heavy nuclei over the wide range of the nuclear chart. We will mainly consider weakly deformed nuclei where the triaxial dynamics plays an important role. The results will be compared with the available experimental data and with the recent studies of quantum phase transitions as well. In Chap. 5, the predictive power of the method presented in Chap. 2 is examined by comparing the spectroscopic properties resulting from the IBM Hamiltonian derived from a relativistic EDF with those obtained from the five-dimensional collective Hamiltonian based on the same EDF. Chapter 6 addresses the question of whether a non-axial nucleus is γ-rigid or unstable, presents a robust regularity of the γ-soft nuclei and how it is realized from a microscopic calculation. The result points to the most suitable IBM description of the γ-soft systems. Chapter 7 discusses the impact of the quantal-mechanical correlation energy on the measurable ground-state properties as an implication of the structural evolution. Chapter 8 is devoted to summary and outlook for possible future research directions. Figure 1.2 indicates how this thesis is organized, and the goal and the motivation of each chapter. For readers' convenience, each chapter and/or section contains an introduction as well as a brief summary. Special attention has been paid so as to clarify the interrelationship among chapters and sections in order to describe a variety of topics in a unified way.

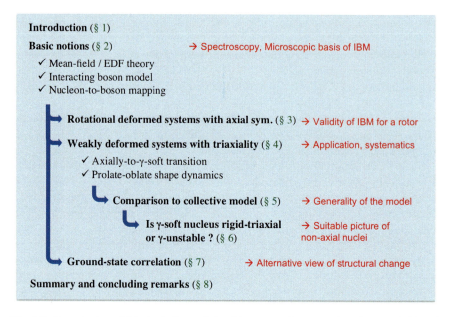

Fig. 1.2 Organization of this thesis. Interrelationship among chapters and sections are indicated. The motivation of each chapter and the outcome drawn from there are indicated on the *right-hand side*

Chapters 2–6, Sect. 7.1, and the Appendices A–B in this thesis are based on the author's original works published already [73, 75, 78, 108–112, 116, 117], coauthoring with (in alphabetical order) M. Albers (Argonne National Laboratory), L. Guo (Graduate University of Chinese Academy of Sciences), T. Nikšić (University of Zagreb), T. Otsuka (University of Tokyo), Zs. Podolyák (University of Surrey), P. H. Regan (University of Surrey), L. M. Robledo (Universidad Autónoma de Madrid), R. Rodríguez-Guzmán (Rice University), P. Sarriguren (Consejo Superior de Investigaciones Científicas, Madrid), N. Shimizu (University of Tokyo), P. D. Stevenson (University of Surrey), and D. Vretenar (University of Zagreb).

References

1. Mayer MG (1949) On closed shells in nuclei II. Phys Rev 75:1969
2. Haxel O, Jensen JHD, Suess HE (1949) On the "magic numbers" in nuclear structure. Phys Rev 75:1766
3. Rainwater J (1950) Nuclear energy level argument for a spheroidal model. Phys Rev 79:432
4. Bohr A, Mottelson BR (1969) Nuclear structure, vol I: Single-particle motion. Benjamin, New York
5. Bohr A, Mottelson BR (1975) Nuclear structure, vol II: Nuclear deformations. Benjamin, New York

References

6. Bohr A (1952) The coupling of nuclear surface oscillations to the motion of individual nucleons. Mat Fys Medd Dan Vid Selsk 26(14):1
7. Bohr A, Mottelson BR (1953) Collective and individual-particle aspects of nuclear structure. Mat Fys Medd Dan Vid Selsk 27(16):1
8. Hill DL, Wheeler JA (1953) Nuclear constitution and the interpretation of fission phenomena. Phys Rev 89:1102
9. Nambu Y, Jona-Lasinio G (1961) Dynamical model of elementary particles based on an analogy with superconductivity, I. II. Phys Rev 122:345
10. Nambu Y, Jona-Lasinio G (1961) Dynamical model of elementary particles based on an analogy with superconductivity, I. II. Phys Rev 124:246
11. Talmi I (1993) Simple models of complex nuclei: the shell model and interacting boson model. Harwood Academic, New York
12. de Shalit A, Talmi I (1963) Nuclear shell theory. Academic Press, New York
13. Greiner W, Maruhn JA (1996) Nuclear models. Springer, Berlin
14. Ring P, Schuck P (1980) The nuclear many-body problem. Springer, Berlin
15. Bender M, Heenen P-H, Reinhard P-G (2003) Self-consistent mean-field models for nuclear structure. Rev Mod Phys 75:121–180
16. Arima A, Iachello F (1975) Collective nuclear states as representations of a SU(6) group. Phys Rev Lett 35:1069
17. Iachello F, Arima A (1987) The interacting boson model. Cambridge University Press, Cambridge
18. Casten RF (2006) Shape phase transitions and critical-point phenomena in atomic nuclei. Nat Phys 2:811
19. Cejnar P, Casten RF, Jolie J (2010) Quantum Phase transitions in the shapes of atomic nuclei. Rev Mod Phys 82:2155
20. Casten RF (1990) Nuclear structure from a simple perspective. Oxford University Press, Oxford
21. Skyrme THR (1959) The effective nuclear potential. Nucl Phys 9:615
22. Vautherin D, Vénéroni M (1969) A Hartree-Fock calculation of 208Pb in coordinate space. Phys Lett B 29:203
23. Vautherin D, Brink DM (1972) Hartree-Fock calculations with Skyrme's interaction. I. Spherical nuclei. Phys Rev C 5:626
24. Erler J, Klüpfel P, Reinhard P-G (2011) Self-consistent nuclear mean-field models: example Skyrme-Hartree-Fock. J Phys G: Nucl Part Phys 38:033101
25. Decharge J, Girod M, Gogny D (1975) Self consistent calculations and quadrupole moments of even Sm isotopes. Phys Lett B 55:361
26. Dechargé J, Gogny D (1980) Hartree-Fock-Bogolyubov calculations with the D1 effective interaction on spherical nuclei. Phys Rev C 21:1568
27. Walecka JD (1974) A theory of highly condensed matter. Ann Phys 83:491
28. Vretenar D, Afanasjev AV, Lalazissis GA, Ring P (2005) Relativistic Hartree-Bogoliubov theory: static and dynamic aspects of exotic nuclear structure. Phys Rep 409:101
29. Nikšić T, Vretenar D, Ring P (2011) Relativistic nuclear energy density functionals: mean-field and beyond. Prog Part Nucl Phys 66:519
30. Hill DL, Wheeler JA (1953) Nuclear constitution and the interpretation of fission phenomena. Phys Rev 102:311
31. Griffin JJ, Wheeler JA (1957) Collective motions in nuclei by the method of generator coordinates. Phys Rev 108:311
32. Bender M, Bertsch G, Heenen P-H (2006) Global study of quadrupole correlation effects. Phys Rev C 73:034322
33. Bender M, Heenen P-H (2008) Configuration mixing of angular-momentum and particle-number projected triaxial Hartree-Fock-Bogoliubov states using the Skyrme energy density functional. Phys Rev C 78:024309
34. Rodríguez-Guzmán R, Egido JL, Robledo LM (2002) Correlations beyond the mean field in magnesium isotopes: angular momentum projection and configuration mixing. Nucl Phys A 709:201

35. Rodriguez-Guzman RR, Egido JL, Robledo LM (2004) Beyond mean field description of shape coexistence in neutron-deficient Pb isotopes. Phys Rev C 69:054319
36. Rodríguez TR, Egido JL (2010) Triaxial angular momentum projection and configuration mixing calculations with the Gogny force. Phys Rev C 81:064323
37. Nikšić T, Vretenar D, Lalazissis GA, Ring P (2007) Microscopic description of nuclear quantum phase transitions. Phys Rev Lett 99:092502
38. Yao JM, Mei H, Chen H, Meng J, Ring P, Vretenar D (2011) Configuration mixing of angular-momentum-projected triaxial relativistic mean-field wave functions. II. Microscopic analysis of low-lying states in magnesium isotopes. Phys Rev C 83:014308.
39. Bonche P, Dobaczewski J, Flocard H, Heenen P-H, Meyer J (1990) Analysis of the generator coordinate method in a study of shape isomerism in ^{194}Hg. Nucl Phys A 510:466
40. Delaroche J-P, Girod M, Libert L, Goutte H, Hilaire S, Peru S, Pillet N, Bertsch GF (2010) Structure of even-even nuclei using a mapped collective Hamiltonian and the D1S Gogny interaction. Phys Rev C 81:014303
41. Nikšić T, Li ZP, Vretenar D, Próchniak L, Meng J, Lalazissis GA, Ring P (2009) Beyond the relativistic mean-field approximation. III. Collective Hamiltonian in five dimensions. Phys Rev C 79:034303.
42. Li ZP, Nikšić T, Vretenar D, Meng J, Lalazissis GA, Ring P (2009) Microscopic analysis of nuclear quantum phase transitions in the $N \approx 90$ region. Phys Rev C 79:054301
43. Li ZP, Nikšić T, Vretenar D, Meng J (2010) Microscopic description of spherical to γ-soft shape transitions in Ba and Xe nuclei. Phys Rev C 81:034316
44. Duguet T, Bender M, Bonche P, Heenen P-H (2003) Shape coexistence in ^{186}Pb: beyond-mean-field description by configuration mixing of symmetry restored wave functions. Phys Lett B 559:201
45. Bender M, Bonche P, Duguet T, Heenen P-H (2004) Configuration mixing of angular momentum projected self-consistent mean-field states for neutron-deficient Pb isotopes. Phys Rev C 69:064303
46. Nikšić T, Vretenar D, Ring P (2006) Beyond the relativistic mean-field approximation. II. Configuration mixing of mean-field wave functions projected on angular momentum and particle number. Phys Rev C 74:064309.
47. Rodríguez TR, Egido JL (2008) A beyond mean field analysis of the shape transition in the Neodymium isotopes. Phys Lett B 663:663
48. Anguiano M, Egido JL, Robledo LM (2001) Particle number projection with effective forces. Nucl Phys A 696:467
49. Dobaczewski J, Stoitsov M, Nazarewicz W, Reinhard P-G (2007) Particle-number projection and the density functional theory. Phys Rev C 76:054315
50. Lacroix D, Duguet T, Bender M (2009) Configuration mixing within the energy density functional formalism: removing spurious contributions from non-diagonal energy kernels. Phys Rev C 79:044318
51. Bender M, Duguet T, Lacroix D (2009) Particle-number restoration within the energy density functional formalism. Phys Rev C 79:044319
52. Duguet T, Bender M, Bennaceur K, Lacroix D, Lesinski T (2009) Particle-number restoration within the energy density functional formalism: are terms depending on non-integer powers of the density matrices viable? Phys Rev C 79:044320
53. Otsuka T, Arima A, Iachello F, Talmi I (1978) Shell model description of interacting bosons. Phys Lett B 76:139
54. Otsuka T, Arima A, Iachello F (1978) Shell model description of interacting bosons. Nucl Phys A 309:1
55. Arima A, Iachello F (1976) Interacting boson model of collective states: I. The vibrational limit. Ann Phys 99:253–317
56. Arima A, Iachello F (1978) Interacting boson model of collective states: II. The rotational limit. Ann Phys 111:201–238
57. Arima A, Iachello F (1979) Interacting boson model of collective states: IV. The O(6) limit. Ann Phys 123:468–492

References

58. Frank A, Van Isacker P (1994) Algebraic methods in molecular and nuclear structure physics. Willey, New York
59. Iachello F, Levine RD (1995) Algebraic theory of molecules. Oxford University Press, Oxford
60. Otsuka T (1993) Microscopic Basis and Introduction to IBM-2. In: Casten RF (ed) Algebraic approaches to nuclear structure. Harwood, Chur, p 195
61. Gambhir YK, Ring P, Schuck P (1982) Microscopic determination of the interacting boson model parameters. Phys Rev C 25:2858
62. Mizusaki T, Otsuka T (1996) Microscopic calculations for O(6) nuclei by the interacting boson model. Prog Theor Phys Suppl 125:97–150
63. Deleze M, Drissi S, Kern J, Tercier TA, Vorlet JP, Rikovska J, Otsuka T, Judge S, Williams A (1993) Systematic study of the mixed ground-state and "intruder" bands in ^{110,112,114}Cd. Nucl Phys A551:269–294
64. Allaart K, Bonsignori G, Savoia M, Paar V (1986) Construction of microscopic boson states and their relevance for IBM-2. Nucl Phys A 458:412
65. Scholten O (1983) Microscopic calculations for the interacting boson model. Phys Rev C 28:1783
66. Otsuka T (1984) "Independent-pair" property of condensed coherent pairs and derivation of the IBM quadrupole operator. Phys Lett B 138:1
67. Otsuka T, Yoshinaga N (1986) Fermion-boson mapping for deformed nuclei. Phys Lett B 168:1
68. Puddu G, Scholten O, Otsuka T (1980) Collective quadrupole states of Xe, Ba and Ce in the interacting boson model. Nucl Phys 348:109–124
69. Van Isacker P, Puddu G (1980) The Ru and Pd isotopes in the proton-neutron interacting boson model. Nucl Phys A 348:125
70. Kaup U, Gelberg A (1979) Description of even-even Krypton sotopes by the interacting boson approximation. Z Phys A 293:311
71. Duval P, Barrett BR (1981) Interacting boson approximation model of the tungsten isotopes. Phys Rev C 23:492
72. Bijker R, Dieperink AEL, Scholten O, Spanhoff R (1980) Description of the Pt and Os isotopes in the interacting boson model. Nucl Phys A 344:207
73. Nomura K, Shimizu N, Otsuka T (2008) Mean-field derivation of the interacting boson model Hamiltonian and exotic nuclei. Phys Rev Lett 101:142501
74. Ginocchio JN, Kirson M (1980) An intrinsic state for the interacting boson model and its relationship to the Bohr-Mottelson model. Nucl Phys A 350:31
75. Nomura K, Shimizu N, Otsuka T (2010) Formulating the interacting boson model by mean-field methods. Phys Rev C 81:044307
76. Kaiser G (1994) A friendly guide to wavelets. Birkhäser, Boston
77. Bohr A, Mottelson BR (1980) Features of nuclear deformations produced by the alignment of individual particles or pairs. Phys Scripta 22:468
78. Nomura K, Otsuka T, Shimizu N, Guo L (2011) Microscopic formulation of the interacting boson model for rotational nuclei. Phys Rev C 83:041302(R).
79. Iachello F (2001) Analytic description of critical point nuclei in a spherical-axially deformed shape phase transition. Phys Rev Lett 87:052501
80. Casten RF, Zamfir NV (2001) Empirical realization of a critical point description in atomic nuclei. Phys Rev Lett 87:052503
81. Iachello F (2000) Dynamical symmetries at the critical point. Phys Rev Lett 85:3580
82. Casten RF, Zamfir NV (2000) Evidence for a possible E(5) symmetry in ^{134}Ba. Phys Rev Lett 85:3584
83. Rodríguez-Guzmán R, Sarriguren P, Robledo LM, Perez-Martín S (2000) Charge radii and structural evolution in Sr, Zr, and Mo isotopes. Phys Lett B 691:202
84. Rodríguez-Guzmán R, Sarriguren P, Robledo LM (2010) Systematics of one-quasiparticle configurations in neutron-rich odd Sr, Zr, and Mo isotopes with the Gogny energy density functional. Phys Rev C 82:044318

85. Rodríguez-Guzmán R, Sarriguren P, Robledo LM (2010) Signatures of shape transitions in odd-A neutron-rich rubidium isotopes. Phys Rev C 82:061302(R).
86. Wood JL, Heyde K, Nazarewicz W, Huyse M, Van Duppen P (1992) Coexistence in even-mass nuclei. Phys Rep 215:101
87. Werner TR, Dobaczewski J, Guidry MW, Nazarewicz W, Sheikh JA (1994) Microscopic aspects of nuclear deformation. Nucl Phys A 578:1
88. Cwiok S, Heenen P-H, Nazarewicz W (2005) Shape coexistence and triaxiality in the superheavy nuclei. Nature 433:705
89. Robledo LM, Rodríguez-Guzmán RR, Sarriguren P (2008) Evolution of nuclear shapes in medium mass isotopes from a microscopic perspective. Phys Rev C 78:034314
90. Sarriguren P, Rodríguez-Guzmán R, Robledo LM (2008) Shape transitions in neutron-rich Yb, Hf, W, Os, and Pt isotopes within a Skyrme-Hartree-Fock + BCS approach. Phys Rev C 77:064322
91. Rodríguez-Guzmán R, Sarriguren P (2007) E(5) and X(5) shape phase transitions within a Skyrme-Hartree-Fock + BCS approach. Phys Rev C 76:064303
92. Egido JL, Robledo LM, Rodríguez-Guzmán RR (2004) Unveiling the origin of shape coexistence in lead isotopes. Phys Rev Lett 93:082502
93. Nazarewicz W (1994) Microscopic origin of nuclear deformations. Nucl Phys A 574:27c
94. Hamamoto I, Mottelson BR (2009) Further examination of prolate-shape dominance in nuclear deformation. Phys Rev C 79:034317
95. Egido JL, Lessing J, Martin V, Robledo LM (1995) On the solution of the Hartree-Fock-Bogoliubov equations by the conjugate gradient method. Nucl Phys A 594:70
96. Robledo LM, Rodríguez-Guzmán R, Sarriguren P (2009) Role of triaxiality in the ground-state shape of neutron-rich Yb, Hf, W, Os and Pt isotopes. J Phys G: Nucl Part Phys. 36:115104
97. Julin R, Helariutta K, Muikku M (2001) Intruder states in very neutron-deficient Hg, Pb and Po nuclei. J Phys G 27:R109
98. Dracoulis GD, Stuchbery AE, Byrne AP, Poletti AR, Polotti SJ, Gerl J, Bark RA (1986) Shape coexistence in very neutron-deficient Pt isotopes. J Phys G 12:L97
99. Dracoulis GD, Fabricius B, Stuchbery AE, Macchiavelli AO, Korten W, Azaiez F, Rubel E, Deleplanque MA, Diamond RM, Stephens FS (1991) Shape coexistence from the structure of the yrast band in 174Pt. Phys Rev C 44:R1246
100. Davidson PM, Dracoulis GD, Kibédi T, Byrne AP, Anderssen SS, Baxter AM, Fabricius B, Lane GJ, Stuchbery AE (1994) Non-yrast states and shape co-existence in ^{172}Os. Nucl Phys A 568:90
101. Davidson PM, Dracoulis GD, Kibédi T, Byrne AP, Anderssen SS, Baxter AM, Fabricius B, Lane GJ, Stuchbery AE (1999) Non-yrast states and shape co-existence in light Pt isotopes. Nucl Phys A A 657:219
102. Kibéti T, Dracoulis GD, Byrne AP, Davidson PM (1994) Low-spin non-yrast states and collective excitations in ^{174}Os, ^{176}Os, ^{178}Os, ^{180}Os, ^{182}Os and ^{184}Os. Nucl Phys A 567:183
103. Kibéti T, Dracoulis GD, Byrne AP, Davidson PM (2001) Low-spin non-yrast states in light tungsten isotopes and the evolution of shape coexistence. Nucl Phys A 688:669
104. Wu CY, Cline D, Czosnyka T, Backlin A, Baktash C, Diamond RM, Dracoulis GD, Hasselgren L, Kluge H, Kotlinski B, Leigh JR, Newton JO, Phillips WR, Sie SH, Srebrny J, Stephens FS (1996) Quadrupole collectivity and shapes of Os-Pt nuclei. Nucl Phys A 607:178
105. Zs Podolyák et al (2000) Isomer spectroscopy of neutron rich ^{190}W$_{116}$. Phys Lett B 491:225
106. Pfützner M et al (2002) Angular momentum population in the fragmentation of ^{208}Pb at 1 GeV/nucleon. Phys Rev C 65:064604
107. Caamaño M et al (2005) Isomers in neutron-rich A $\approx$ 190 nuclides from ^{208}Pb fragmentation. Eur Phys J A 23:201
108. Albers M, Warr N, Nomura K, Blazhev A, Jolie J, Mücher D, Bastin B, Bauer C, Bernards C, Bettermann L, Bildstein V, Butterworth J, Cappellazzo M, Cederkäll J, Cline D, Darby I, Daugas JM, Davinson T, De Witte H, Diriken J, Filipescu D, Fiori E, Fransen C, Gaffney LP, Georgiev G, Gernhäuser R, Hackstein M, Heinze S, Hess H, Huyse M, Jenkins D, Konki J, Kowalczyk M, Kröll T, Lutter R, Marginean N, Mihai C, Moschner K, Napiorkowski P,

Nowak K, Otsuka T, Pakarinen J, Pfeiffer M, Radeck D, Reiter P, Rigby S, Robledo LM, Rodríguez-Guzmán R, Rudigier M, Sarriguren P, Scheck M, Seidlitz M, Siebeck B, Simpson G, Thoele P, Thomas T, Van de Walle J, Van Duppen P, Vermeulen M, Voulot D, Wadsworth R, Wenander F, Wimmer K, Zell KO, Zielinska M (2012) Evidence for a smooth onset of deformation in the neutron-rich Kr isotopes. Phys Rev Lett 108:062701

109. Nomura K, Otsuka T, Rodríguez-Guzmán R, Robledo LM, Sarriguren P (2011) Structural evolution in Pt isotopes with the interacting boson model Hamiltonian derived from the Gogny energy density functional. Phys Rev C 83:014309
110. Nomura K, Otsuka T, Rodríguez-Guzmán R, Robledo LM, Sarriguren P, Regan PH, Stevenson PD, Zs Podolyák (2011) Spectroscopic calculations of the low-lying structure in exotic Os and W isotopes. Phys Rev C 83:051303
111. Nomura K, Otsuka T, Rodríguez-Guzmán R, Robledo LM, Sarriguren P (2011) Collective structural evolution in Yb, Hf, W, Os and Pt isotopes. Phys Rev C 84:054316
112. Nomura K, Nikšić T, Otsuka T, Shimizu N, Vretenar D (2011) Quadrupole collective dynamics from energy density functionals: collective Hamiltonian and the interacting boson model. Phys Rev C 84:014302
113. Nikšić T, Vretenar D, Ring P (2008) Relativistic nuclear energy density functionals: adjusting parameters to binding energies. Phys Rev C 78:034318
114. Davydov AS, Filippov GF (1958) Rotational states in even atomic nuclei. Nucl Phys 8:237
115. Wilets L, Jean M (1956) Surface oscillations in even-even nuclei. Phys Rev 102:788
116. Nomura K, Shimizu N, Vretenar D, Nikšić T, Otsuka T (2012) Robust regularity in γ-soft nuclei and its novel microscopic realization. Phys Rev Lett 108:132501
117. Nomura K (2011) Microscopic derivation of IBM and structural evolution in nuclei. In: AIP conference proceedings, vol 1355, International symposium; New faces of atomic nuclei, pp 23–28.

Chapter 2
Basic Notions

2.1 General Remarks

Before the thorough discussions on each particular case, we first present our principal idea. It is often quite reasonable to start with the microscopic calculation of self-consistent mean-field (potential) energy surface with energy density functional (EDF), from which collective spectra and transition rates are derived (cf. Fig. 2.1). Since the energy surface reflects intrinsic deformation, the question arises here: How can we incorporate the energy-surface calculation into the relevant measurable spectroscopic properties with good symmetries? There have been many EDF-based schemes which derive spectroscopic properties from the energy surface, including the generator coordinate method (GCM), collective Hamiltonian approach, etc.[1] These studies are, however, still computationally quite complicated and demanding. Our motivation to employ the interacting boson model (IBM) is twofold: First, IBM can be utilized as an effective theory to generate excitation energies and transition rates with good symmetries in a computationally much moderate way, in comparison to other mean-field based, spectra-generating approaches mentioned above. The second is rather profound. Formulating the IBM by EDF approach should be of certain interest because it may help clarifying the major long-lasting problem of IBM concerning its microscopic foundation.

Note that this chapter is not intended to the complete review of the mean-field theory and the IBM. For pedagogical literature of these two models, the interested reader is referred to the textbook by Ring and Schuck [1] or more recent review article by Bender et al. [2] for the former, and the textbook by Iachello and Arima [3, 4] for the latter.

---

[1] For instance, the GCM wave function is constructed by the mixing of mean-field states at many different configurations of collective coordinates.

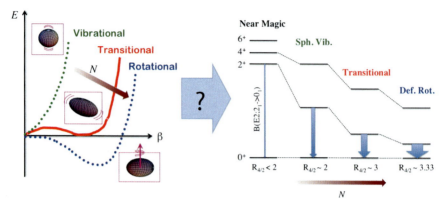

Fig. 2.1 Illustration of the principal idea. *Left-hand side* shows the collective potential energy surface for axially deformed nucleus. With the nucleon number N, the shape of the nucleus changes and the form of the potential changes accordingly. The finding from the energy-surface analysis is subsequently incorporated into the calculation of spectroscopic properties (*right-hand side*)

2.2 Self-Consistent Mean-Field Models

The discussion in this thesis focuses first on self-consistent mean-field models, where one derives in principle a potential for nucleons iteratively, starting from a set of single-nucleon wave functions. This process is apparently at the level of Hartree-Fock approximation, In realistic calculation of nuclei in which the pairing effect plays a crucial role in reproducing various intrinsic properties of finite nuclei, the Hartree-Fock theory must be extended so as to include the pairing field in the particle-particle channel of the Hartree-Fock-Bogoliubov (HFB) equation, which treats both fields simultaneously. Attention has to be paid for the usage of the terminology *mean field* because it refers to the models that involve both self-consistent and pairing fields, and the particle-hole part of the Hamiltonian in the HFB equation. In this thesis the mean-field model stands for both of these. Since the HFB calculation is in general rather computationally demanding, the pairing correlation can be incorporated separately from the Hartree-Fock calculation in the BCS approximation, which is known as the Hartree-Fock plus BCS (HF+BCS) model. The HF+BCS model can be a simplification of the HFB theory and has been also widely used. The basics of the HF+BCS calculation, which is in a sense already well-established, is described in Appendix A.2. In this section, we shall rather explain the density-dependent effective interactions for nuclear many-body system and the constrained mean-field models, which yield the (potential) energy landscape in terms of the geometrical deformation variables and which are the starting point of this work.

2.2.1 Density-Dependent Force

The analysis starts from the constrained mean-field calculations with a given density-dependent effective interaction. The Hartree-Fock theory with the density-dependent force of nuclei can be viewed as a specific kind of the density functional theory (DFT). Originally the DFT was developed by Hohenberg and Kohn [5] to deal with the ground-state of inhomogeneous system of interacting electrons. According to the Hohenberg-Kohn theorem [5], there exists a universal functional of the density for a given system, and the functional in principle gives the exact ground-state energy of the system by the variation with respect to the density. More later the original Hohenberg-Kohn theory was rendered by Kohn and Sham [6] a universal prescription of a given quantum many-body system, leading to the self-consistent equations analogous to the Hartree and the Hartree-Fock equations. In the field of quantum chemistry, the DFT gives very good description of ground-state properties of many-electron systems, and the Coulomb force acting among electrons and between an electron and nucleus is so uniquely-determined compared to effective nuclear force that the DFT has been adopted as a reliable *ab initio* method. In fact, there exist a number of energy density functionals (EDFs) in nuclear physics, and hence the question of what is the most appropriate and universal density functional for nuclear many-body problem remains unanswered and has been one of the open problems in nuclear theory. Nevertheless, thanks to the recent considerable theoretical investigations, one can deal with various nuclear intrinsic properties with remarkable accuracy with a given energy density functional. The popular EDFs in nuclear physics, that are commonly used for the past decades and employed in the calculations presented in this thesis, have been the Skyrme [7–9] and Gogny [10, 11] interactions, and other interactions in the relativistic mean-field framework [12]. The subsequent sections describe these EDFs.

Skyrme Energy Density Functional

The Skyrme interaction or the zero-range effective interaction was originally proposed by Skyrme [7], and more later revisited by Vautherin and Vénéroni [8] and Vautherin and Brink [9] as to formulate it in density-dependent form capable of the universal description as it is today.

Nuclear ground-state energy is obtained by minimizing the following energy functional $\mathscr{E}$, consisting of kinetic energy $\mathscr{E}_{\text{kin}}$, Skyrme energy density functional $\mathscr{E}_{\text{Skyrme}}$, the Coulomb energy $\mathscr{E}_{\text{coul}}$, the pairing functional $\mathscr{E}_{\text{pair}}$, and the correction for spurious center-of-mass motion $\mathscr{E}_{\text{corr}}$:

$$\mathscr{E} = \mathscr{E}_{\text{kin}} + \mathscr{E}_{\text{Skyrme}} + \mathscr{E}_{\text{coul}} + \mathscr{E}_{\text{pair}} - \mathscr{E}_{\text{corr}}. \tag{2.1}$$

All contributions relevant to the many-body effects are represented by the Skyrme functional $\mathscr{E}_{\text{Skyrme}}$ and, since the nucleus is a superconductor, the pairing functional $\mathscr{E}_{\text{pair}}$ should play a role.

The Skyrme-type effective interaction is composed of two- and three-body parts. When one writes the many-body Hamiltonian as

$$\hat{H} = \sum_i \hat{T}_i + \sum_{i<j} \hat{V}_{ij}^{(2)} + \sum_{i<j<k} \hat{V}_{ijk}^{(3)}, \qquad (2.2)$$

the two-body part of the Skyrme force $\hat{V}_{12}^{(2)}$ is given by

$$\hat{V}_{12}^{(2)} = t_0(1 + x_0 \hat{P}_\sigma)\delta(\mathbf{r}_1 - \mathbf{r}_2) + \frac{1}{2}t_1(1 + x_1 \hat{P}_\sigma)[\hat{\mathbf{k}}'^2 \delta(\mathbf{r}_1 - \mathbf{r}_2) + \delta(\mathbf{r}_1 - \mathbf{r}_2)\hat{\mathbf{k}}^2]$$
$$+ t_2(1 + x_2 \hat{P}_\sigma)\hat{\mathbf{k}}'\delta(\mathbf{r}_1 - \mathbf{r}_2)\hat{\mathbf{k}} + i W_0(\hat{\sigma}_1 + \hat{\sigma}_2) \cdot \hat{\mathbf{k}}' \times \delta(\mathbf{r}_1 - \mathbf{r}_2)\hat{\mathbf{k}}, \qquad (2.3)$$

while the three body part $\hat{V}_{123}^{(3)}$ in Eq. (2.2) by

$$\hat{V}_{123}^{(3)} = t_3 \delta(\mathbf{r}_1 - \mathbf{r}_2)\delta(\mathbf{r}_2 - \mathbf{r}_3), \qquad (2.4)$$

which is reduced to a two-body density-dependent term as

$$\hat{V}_{123}^{(3)} = \frac{1}{6}(1 + x_3 \hat{P}_\sigma)\delta(\mathbf{r}_1 - \mathbf{r}_2)\rho^\alpha\left(\frac{\mathbf{r}_1 + \mathbf{r}_2}{2}\right) \qquad (2.5)$$

for spin-saturated even-even nuclei. Notations $\mathbf{r}_1 - \mathbf{r}_2$ and $\hat{P}_\sigma = (1 + \hat{\sigma}_1 \cdot \hat{\sigma}_2)/2$ stand for the relative distance between two nucleons and the operator exchanging spins σ_1 with σ_2, respectively. The last term on the right-hand side (RHS) of Eq. (2.3) represents the spin-orbit interaction. The relative momenta $\hat{\mathbf{k}} = (\nabla_1 - \nabla_2)/2i$ and $\hat{\mathbf{k}}' = -(\nabla_1 - \nabla_2)/2i$ act on a right and a left sides, respectively. Coefficients t_i ($i = 0, 1, 2, 3$), x_0 and W_0 are free parameters which are adjusted phenomenologically to reproduce the bulk properties of finite nuclei in the Skyrme mean-field calculations.

The Hartree-Fock equation can be obtained by evaluating the expectation value of the Hamiltonian with respect to the Hartree-Fock basis, which is normally Slater determinant $|\Phi_{HF}\rangle$ (see Appendix A), as $E = \langle \Phi_{HF}|\hat{H}|\Phi_{HF}\rangle$. To obtain the energy functional, it is more transparent to introduce the following densities,

- Particle-number densities for proton and neutron:

$$\rho_q(\mathbf{r}) = \sum_{k,\sigma} v_{q,k}|\phi_k(\mathbf{r}, \sigma, q)|^2, \qquad (2.6)$$

where $\phi_k(\mathbf{r}, \sigma, q)$ is the single-particle wave function of the state k and the indices of spin $\sigma = \pm 1/2$ and isospin $q = \pm 1/2$ (plus for proton and minus for neutron, or vise versa). $v_{q,k}$ is the BCS occupation factor (for definition, see Appendix A).
- Kinetic energy densities for proton and neutron:

$$\tau_q(\mathbf{r}) = \sum_{k,\sigma} v_{q,k}|\nabla\phi_k(\mathbf{r}, \sigma, q)|^2 \qquad (2.7)$$

2.2 Self-Consistent Mean-Field Models

- Spin-orbit current:

$$\mathbf{J}_q(\mathbf{r}) = -i \sum_{k,\sigma,\sigma'} v_{q,k} \phi_k^*(\mathbf{r}, \sigma, q)[\nabla \phi_k(\mathbf{r}, \sigma', q) \times \langle \sigma | \hat{\sigma} | \sigma' \rangle], \quad (2.8)$$

where $\langle \sigma | \hat{\sigma} | \sigma' \rangle$ is the matrix elements of spin operator.

Using these notations, the Skyrme energy density functional $\mathcal{E}_{\text{Skyrme}}$ can be rewritten by using the Hamiltonian density $\hat{\mathcal{H}}(\mathbf{r})$ as:

$$\mathcal{E}_{\text{Skyrme}} = \int d^3 \mathbf{r} \hat{\mathcal{H}}(\mathbf{r}), \quad (2.9)$$

where $\hat{\mathcal{H}}(\mathbf{r})$ is given by

$$\begin{aligned}\hat{\mathcal{H}}(\mathbf{r}) = {} & \frac{\hbar^2}{2m}\tau + B_1 \rho^2 + B_2(\rho_n^2 + \rho_p^2) \\ & + B_3(\rho\tau - \mathbf{J}^2) + B_4(\rho_n \tau_n - \mathbf{J}_n^2 + \rho_p \tau_p - \mathbf{J}_p^2) \\ & + B_5 \rho \Delta \rho + B_6(\rho_n \Delta \rho_n + \rho_p \Delta \rho_p) + B_7 \rho^{2+\alpha} + B_8 \rho^\alpha (\rho_n^2 + \rho_p^2) \\ & + B_9(\rho \nabla \mathbf{J} + \rho_n \nabla \mathbf{J}_n + \rho_n \nabla \mathbf{J}_n) \end{aligned} \quad (2.10)$$

with $\rho = \rho_p + \rho_n$ and $\mathbf{J} = \mathbf{J}_n + \mathbf{J}_p$ representing the total (scalar) density and current, respectively. The parameters Bs in Eq. (2.10) are the combination of the Skyrme parameters in the original notations in Eq. (2.3).

It should be noted that the Skyrme energy density functional $\mathcal{E}_{\text{Skyrme}}$ is written also as

$$\mathcal{E}_{\text{Skyrme}} = \sum_{T=0,1} \left(\mathcal{E}_T^{(\text{even})} + \mathcal{E}_T^{(\text{odd})} \right), \quad (2.11)$$

where $\mathcal{E}_T^{(\text{even})}$ and $\mathcal{E}_T^{(\text{odd})}$ on the RHS represent the components constructed from time-even and the time-odd densities, respectively. The sum is over the isospin channel $T = 0, 1$. The time-even component $\mathcal{E}^{(\text{even})}$ is the only part which contributes to the static mean-field calculations of ground-state of even-even nuclei. The time-odd component $\mathcal{E}^{(\text{odd})}$ is needed for rotating odd-mass and odd-odd systems with relatively high spin so that the Skyrme functional becomes time-reversal invariant.

When the pairing correlation is taken into account, the energy functional is formulated in the Hartree-Fock-Bogoliubov model, which introduces the quasi-particle wave functions and gives rise to the pairing functional for the particle-particle channel aside from the Skyrme functional for particle-hole channel.

Gogny Energy Density Functional

Because of its zero-range nature, it appears that the Skyrme interaction does not fully simulate the long-range part of the realistic effective nucleon-nucleon interaction. The finite-range force was originally proposed by Brink and Boeker [13] but, in order to reproduce the nuclear binding energies precisely, one needs to modify the interaction so as to include density-dependent term and spin-orbit force. Gogny introduced alternative effective interaction that has a similar structure to the Skyrme interaction, so that t_0, t_1 and t_2 components of the Skyrme interaction of Eq. (2.3) are replaced by a sum of two Gaussian combined with spin and isospin exchange [14]:

$$V_{12} = \sum_{i=1,2} e^{-(\mathbf{r}_1-\mathbf{r}_2)^2/\mu_i^2} (W_i + B_i \hat{P}_\sigma - H_i \hat{P}_\tau - M_i \hat{P}_\sigma \hat{P}_\tau)$$
$$+ i W_0 (\hat{\sigma}_1 + \hat{\sigma}_2) \cdot \hat{\mathbf{k}}' \times \delta(\mathbf{r}_1 - \mathbf{r}_2) \hat{\mathbf{k}} + t_3 (1 + x_3 \hat{P}_\sigma) \delta(\mathbf{r}_1 - \mathbf{r}_2) \rho^\alpha \left(\frac{\mathbf{r}_1 + \mathbf{r}_2}{2} \right). \quad (2.12)$$

The parameters in Eq. (2.12) are determined as usual by fitting to the experimental data for the properties of finite nuclei.

Contrary to the Skyrme interaction, which has more than a hundred of different types of parameterizations, the Gogny-type interaction has much less. One of the most popular interactions is the parametrization D1S [15], whose predictive power has been already shown to be valid when applied to the description of not only the intrinsic properties of a nucleus but also the spectroscopic analyses in any mass region of the nuclear chart. More recently new types of the Gogny interactions have been proposed: the D1N force [16], which is more oriented to the astrophysical interest, and the D1M force [17], which is derived based on the spectroscopic calculations in terms of the five-dimensional collective Hamiltonian approach taking into account necessary correlation energies for the quadrupole degrees of freedom and which is also the first Gogny model associated with nuclear masses. Here we note some studies to design a new Gogny interaction so as to include the tensor force explicitly (e.g., Ref. [18]). These studies addressed the computational problems that generally occur in the number-projected mean-field calculation, and were further aimed at the structure analyses of exotic nuclei as well.

Observables for finite nuclei are obtained in a similar way to the case of Skyrme energy density functional.

Relativistic Energy Density Functionals

Other energy density functional model completely different from the Skyrme and Gogny type interactions consists in the relativistic mean-field (RMF) framework. Contrary to the non-relativistic framework, the RMF is more connected with the sub-nuclear degrees of freedom through the meson fields, and as such is formulated in a more fundamental field theoretical way. The density functional approach based

2.2 Self-Consistent Mean-Field Models

on the RMF has been successful to almost similar extent to Skyrme and Gogny interactions [19, 20].

Another unique features of the nucleus among other quantal systems can be a strong spin-orbit interaction. A benefit brought about by the use of RMF framework is such that the spin-orbit force emerges naturally there and can be included automatically in the effective Lagrangian. Thus the number of parameters is relatively small in the RMF framework as compared to the non-relativistic EDFs. Note that, however, the exchange (Fock) term in the Hartree-Fock equation is often neglected in many RMF calculations. Its effect is included only implicitly in the parameters of the RMF density functional, that are fixed by the phenomenological fit.

With similar treatment of the pairing correlation, Coulomb effect and the center-of-mass correction to the non-relativistic density functionals (cf. Eq. (2.1)), the relativistic energy density functional $\mathcal{E}_{\mathrm{RMF}}$ can be generally given as (when $\hbar = 1$ and $c = 1$ are assumed as usual)

$$\mathcal{E}_{\mathrm{RMF}} = \mathcal{E}_{\mathrm{nucl}} + \mathcal{E}_{\mathrm{meson}} + \mathcal{E}_{\mathrm{coupl}} + \mathcal{E}_{\mathrm{nonl}}, \tag{2.13}$$

The term $\mathcal{E}_{\mathrm{nucl}}$ is associated with the Dirac equation for a single nucleon and is given as

$$\mathcal{E}_{\mathrm{nucl}} = \sum_{k=1}^{\mathrm{occ}} v_k^2 \bar{\psi}_k (-i\gamma \cdot \nabla + m_B) \psi_k, \tag{2.14}$$

where v_k^2 denotes the BCS occupation probability for the state k, which runs through all occupied states. The second term $\mathcal{E}_{\mathrm{meson}}$ is deduced from the Klein-Gordon equations for the meson fields Ψ_M with M denoting the σ, ω and ρ mesons, and is written as

$$\mathcal{E}_{\mathrm{meson}} = \sum_{M=\sigma,\omega,\rho} \frac{1}{2} \bar{\Psi}_M (-\Delta + m_M) \Psi_M. \tag{2.15}$$

For the rest, the coupling between nucleon and meson fields, $\mathcal{E}_{\mathrm{coupl}}$, and the non-linear terms $\mathcal{E}_{\mathrm{nonl}}$ arising from the self-coupling among meson fields:

$$\mathcal{E}_{\mathrm{coupl}} = g_\sigma \Psi_\sigma \rho_{s0} + g_\omega \Psi_\omega^\mu \rho_{\mu 0} + g_\rho \Psi_\rho^\mu \rho_{\mu,1}$$

$$\mathcal{E}_{\mathrm{nonl}} = \frac{1}{3} b_2 \Psi_\sigma^3 + \frac{1}{4} b_3 \Psi_\sigma^4 + \frac{1}{4} c_3 (\Psi_{\omega,\mu} \Psi_\omega^\mu), \tag{2.16}$$

where $g_\sigma, g_\omega, g_\rho, b_2, b_3$ and c_3 are parameters that are fixed by the phenomenological fit, and μ for Ψ's represents an index for the relativistic four vector. ρ_{sT} and $\rho_{\mu T}$ are scalar and vector densities, respectively.

2.2.2 Constrained Mean Field

To interpret the geometrical shape of the nuclei in the ground state, the energy landscape (energy surface) with multipole degrees of freedom, such as of quadrupole and octupole types, are often considered. As a representative for the self-consistent mean-field calculation to obtain the energy surface, specifically the one for the quadrupole deformation, we here describe one of the most basic frameworks which are widely used in the market: the constrained Skyrme Hartree-Fock plus BCS method for axial and triaxial degrees of freedom in any form of the single-particle state [1] like the coordinate-space [21, 22] and the harmonic oscillator [23] representations. Pairing correlation is taken into account in the BCS approximation [24, 25] in addition to the Hartree-Fock solution. Some more details can be found in Appendix A.

The code ev8 [21], developed by Bonche et al., is one of the well-organized, user-friendly computer program for performing the self-consistent constrained mean-field calculation in the HF+BCS approach using Skyrme density functional. In many of the author's works the ev8 code was used for calculating the constrained energy surface, and the following discussion treats mainly this ev8 code. In the ev8 code the single-particle wave functions are represented by a three-dimensional Cartesian grid and are assumed to be symmetric with respect to xyz planes [21, 22]. We use a mesh spacing of 0.8 fm throughout, which is good enough to give accurate mean-field solution [21]. Mean-field equations are solved iteratively, where the single-particle wave function at a given step of the iteration is provided by the imaginary time method [21]. BCS equation is solved subsequently by using the eigenvectors at the corresponding step of the iteration. Also the two-body center-of-mass motion is subtracted from the mean-field solution at the final step of the iteration. We use the Skyrme SLy4 [26] and SkM* [27] interactions, while the following results do not depend too much on the choice of Skyrme parameterizations as long as the usual ones are taken.

In performing the HF+BCS calculation, we employ the density-dependent zero-range type of the pairing interaction, which is truncated both below and above the Fermi surface by 5 MeV for both protons and neutrons. The pairing force in the functional form, denoted by $\mathscr{E}_{\text{pair}}$, is written as

$$\mathscr{E}_{\text{pair}} = \frac{1}{4} \sum_{q=\pm 1/2} V_q \int d^3\mathbf{r} \left[1 - \frac{\rho(r)}{\rho_c}\right] \tilde{\rho}_q(\mathbf{r}) \tilde{\rho}_q^*(\mathbf{r}), \quad (2.17)$$

where $\tilde{\rho}_q(\mathbf{r})$ is defined as $\tilde{\rho}_q(\mathbf{r}) = \sum_{k,\sigma} u_{q,k} v_{q,k} |\phi_k(\mathbf{r}, \sigma, q)|^2$. The factor $u_{k,q}$ in Eq. (2.17) represents non-occupation amplitude and is connected to the $v_{k,q}$ factor for state k as $u_{k,q}^2 + v_{k,q}^2 = 1$ for each of the proton ($q = +1/2$) and neutron ($q = -1/2$) (cf. Appendix A.1). ρ_c stands for the density at the nuclear surface and is fixed as $\rho_c = 0.16\,\text{fm}^{-3}$, being consistent with the nuclear matter properties. The fixed value of the pairing strength $V_q = -1250\,\text{MeV}\,\text{fm}^3$ for both proton ($q = +1/2$) and neutron ($q = -1/2$) is taken. While the use of this value is recommended in the literature, it is not obvious that one can use the common value

2.2 Self-Consistent Mean-Field Models

of the pairing strength for different mass regions. Despite this potential ambiguity of the BCS pairing strength, it is of little importance here to examine each individual case employing the pairing strength different from 1250 MeV fm^3, as indeed it can be shown that the final spectroscopic results do not depend on the every detail of the pairing properties. This will be confirmed in later chapters in which more self-consistent, Hartree-Fock-Bogoliubov calculation, which include both the self-consistent field and the pairing field simultaneously, is employed. Concerning the pairing channel, one could in principle consider both $T=0$ and $T=1$ components. However, since in many of the medium-heavy and heavy nuclei the protons and neutrons occupy different major shells, pairing correlations between identical particles dominate over the proton-neutron pairing. Indeed, many of the earlier works employ only the pairing correlations between identical particles of $T=1$ component. The work presented in this thesis also adopt only the pairing correlation between identical particles of the $T=1$ channel. For lighter nuclei with $N \approx Z$, proton-neutron pairing in both $T=0$ and $T=1$ channels may become stronger.

Instead of the exact variation-after-projection treatment of the particle number, we employ the Lipkin-Nogami prescription [28–30], which imposes the additional constraint proportional to $\langle \hat{N}^2 \rangle$ with $\hat{N}$ being the number operator.

It should be noted that, apart from the fact that many of the current density-dependent interactions are constructed by the fit to the experimental data of ground-state properties, the procedure to obtain the spectroscopic observables mentioned in the following is fully microscopic without any adjustable parameters.

The quadratic constraint is imposed on the Skyrme mean-field calculation in such a way as to add the potential term $\sum_{m=0,2} C_m (\langle \hat{Q}_{2m} \rangle - \mu_m)^2$ to the Hamiltonian.[2] Here, $\langle \hat{Q}_{2m} \rangle$ stands for the expectation value of a component of the mass quadrupole moment $\hat{Q}_{2m} = r^2 Y_{2m}$. Here C_m and μ_m represent the strength of the constraint and some desired value for the quadrupole deformation of interest, respectively. Energy surface is then a function of $\langle \hat{Q}_{20} \rangle$ and $\langle \hat{Q}_{22} \rangle$. It is more convenient to draw the energy surface in terms of the geometrical deformation variables β and γ [36] defined as

$$\beta = \sqrt{\frac{5}{16\pi} \frac{4\pi}{3}} \frac{1}{AR_0^2} q_0 \quad \text{and} \quad \gamma = \tan^{-1}\left(\sqrt{2}\frac{\langle \hat{Q}_{22} \rangle}{\langle \hat{Q}_{20} \rangle}\right), \quad (2.18)$$

which represent axially symmetric deformation and the triaxiality, respectively, Here R_0, with the parameter $r_0 = 1.2$ fm, and $q_0 \equiv \sqrt{\langle \hat{Q}_{20} \rangle^2 + 2\langle \hat{Q}_{22} \rangle^2}$ stand for the empirical nuclear radius and the (mass) quadrupole moment, respectively. Note that it is sufficient to consider the problem in the range $0° \leq \gamma \leq 60°$ since, in the quadrupole deformation, the nuclear shape remains unchanged under the interchange of all three axes of the intrinsic frame.

---

[2] Other technical aspects of the constrained Hartree-Fock calculations are described in Appendix A.3.

2.3 Interacting Boson Model

Let us now review the basic features of the interacting boson model [3, 4], which is generally referred to as IBM in its short-hand notation. The building blocks of the IBM is the monopole s and the quadrupole d bosons, reflecting the collective $L^\pi = 0^+$ and 2^+ collective pairs of valence nucleons. The number of bosons is half the number of valence nucleons counted from the nearest closed shells [31, 32], and does not distinguish between hole and particle configurations. With this picture and the basic interactions between these bosons can one simulate the nuclear collective spectra.

In this thesis we consider the proton-neutron interacting boson model (IBM-2) [31, 32], which distinguishes between proton and neutron bosons. The IBM-2 is more closely connected with a microscopic picture than the original version of the IBM (so-called IBM-1), which does not make distinction between proton and neutron degrees of freedom. The IBM-2 is comprised of the so-called proton (neutron) s_π (s_ν) and the proton (neutron) d_π (d_ν) bosons, which reflect collective $L = 0^+$ and 2^+ proton (neutron) pairs, respectively [31, 32]. The number of proton (neutron) bosons N_ρ is equal to half the numbers of valence protons (neutrons) [31, 32]. Since in medium-heavy and heavy nuclei protons and neutrons occupy in the different major shells, one has only to consider proton-proton and neutron-neutron pairs in the IBM-2 framework, neglecting the proton-neutron pair.

To make the discussion as simple as possible, we first consider the simplest (original) version of the IBM (IBM-1).

2.3.1 Algebras

Exactly-solvable models often give profound insight into a system of interest. The IBM takes on a feature belonging to them: algebraic aspect called dynamical symmetry.

Since the d boson has five components $d_{\pm 2}$, $d_{\pm 1}$ and d_0, one has in total six bosons. Let us write these bosons as

$$b_{0,0} = s, \quad b_{2,2} = d_{+2}, \quad b_{2,1} = d_{+1}, \quad b_{2,0} = d_0, \quad b_{2,-1} = d_{-1}, \quad b_{2,-2} = d_{-2}, \tag{2.19}$$

which satisfies boson commutation relations $[b_{lm}, b^\dagger_{l'm'}] = \delta_{ll'}\delta_{mm'}$, then there are 36 bilinear products of boson creation and annihilation operators written in a coupled form as

$$G^{(k)}_\kappa(l, l') = [b^\dagger_l \times \tilde{b}_{l'}]^{(k)}_\kappa, \tag{2.20}$$

where $l, l' = 0$ or 2 and the notation $\tilde{b}_{lm} = (-1)^{l+m} b_{l,-m}$ is introduced to conserve the rotational invariance and k and κ represent the rank of the tensor product and its projection, respectively. Note that, however, the indices m and m' are omitted

2.3 Interacting Boson Model

in Eq. (2.20) as they are irrelevant to this expression. Since $G_\kappa^{(k)}(l,l')$ satisfies the commutation relation in the closed form[3]:

$$[G_\kappa^{(k)}(l,l'), G_{\kappa'}^{(k')}(l'',l''')] = \sum_{k'',\kappa''} \sqrt{(2k+1)(2k'+1)}(k\kappa k'\kappa'|k''\kappa'')(-1)^{k-k'}$$

$$\times \left[(-1)^{k+k'+k''} \begin{Bmatrix} k & k' & k'' \\ l''' & l & l' \end{Bmatrix} \delta_{l'l''} G_{\kappa''}^{(k'')}(l,l''')\right.$$

$$\left. - \begin{Bmatrix} k & k' & k'' \\ l'' & l' & l \end{Bmatrix} \delta_{l,l'''} G_{\kappa''}^{(k'')}(l'',l')\right], \quad (2.21)$$

the 36 operators $G_\kappa^{(k)}(l,l')$ are the generators for U(6) (or SU(6), precisely speaking[4]) group. $(k\kappa k'\kappa'|k''\kappa'')$ and the curly bracket {} on the RHS of Eq. (2.21) mean the Clebsch-Gordan coefficient (or $3-j$ symbol) and the Wigner $6-j$ symbol, respectively. It is proved that the U(6) symmetry is broken into some chains of its subgroups, depending on the different combinations of the generators. Since the nucleus has the rotational invariance (O(3) symmetry), there are only three possibilities for the subgroups: U(5), SU(3) and O(6) symmetries. When an operator is commutable with all the generators of a given group, then the operator is called Casimir operator of the group. For instance, total angular momentum squared, $\hat{L}^2$, is a Casimir operator for the O(3) group, which commutes with all the generators of the O(3) group, i.e., $\hat{L}_1$, $\hat{L}_2$ and $\hat{L}_3$. When the Hamiltonian is written as a linear combination of the Casimir operators of subsets of a group G,

$$\hat{H} = a\hat{C}_G + b\hat{C}_{G'} + c\hat{C}_{G''} + \cdots \quad (2.22)$$

with

$$G \supset G' \supset G'' \cdots, \quad (2.23)$$

where $\hat{C}_G$, $\hat{C}_{G'}$ ... are the Casimir operators of group G, G', ..., respectively, then the system of interest possesses the dynamical symmetry and the eigenvalue problem is solved analytically. This gives rise to the analytical form of the eigenvalues and the wave functions. The coefficients a, b, c, ..., in Eq. (2.22) are determined normally from the fit to the experimental data. Having a general boson Hamiltonian, dynamical symmetry is realized when the coefficients of the Hamiltonian take specific (for that

---

[3] The product of k_1th and k_2th tensor operators, $\hat{T}^{(k_1)}$ and $\hat{T}^{(k_2)}$, is given by $[\hat{T}^{(k_1)} \times \hat{T}^{(k_2)}]_\kappa^{(k)} = \sum_{\kappa_1,\kappa_2}(k_1k_2\kappa_1\kappa_2|k\kappa)\hat{T}_{\kappa_1}^{(k_1)}\hat{T}_{\kappa_2}^{(k_2)}$, where $(k_1k_2\kappa_1\kappa_2|k\kappa)$ is the Clebsch-Gordan coefficient and $\kappa = \kappa_1 + \kappa_2$ is satisfied. The scalar product ($k=0$) of lth tensor operators $U^{(l)} \cdot V^{(l)}$ is given as $\hat{U}^{(l)} \cdot \hat{V}^{(l)} = (-1)^l\sqrt{2l+1}[\hat{U}^{(l)} \times \hat{V}^{(l)}]_0^{(0)}$, which leads to a simpler form $\hat{U}^{(l)} \cdot \hat{V}^{(l)} = \sum_\kappa (-1)^\kappa \hat{U}_\kappa^{(l)} \hat{V}_{-\kappa}^{(l)}$.

[4] In the language of IBM, the labels $U(N)$ and $O(N)$ are used rather than the labels $SU(N)$ and $SO(N)$, respectively, and hence U(N) (O(N)) always means SU(N) (SO(N)) unless otherwise specified.

reason, the symmetry emerges *dynamically*) values. Therefore, one could identify, depending on these coefficients, which symmetry is realized in a system as well as the extent to which the symmetry is broken (or, mixed).

To list the three dynamical symmetries:

U(5) Limit

The first subgroup U(5) appears as

$$\text{(chain I)} \quad U(6) \supset U(5) \supset O(5) \supset O(3) \supset O(2), \tag{2.24}$$

where s and d bosons are completely decoupled in the U(5) subgroup. The quantum numbers of the U(5) are n_d (the number of d bosons), ν (d-boson seniority), n_Δ (number of triple bosons coupled to $J^+ = 0^+$), L (total angular momentum) and M (z component of L in the laboratory frame). The wave function and the eigenvalue are specified by these quantum numbers. Eigenvalues are given by

$$E(n_d, \nu, n_\Delta, L, M) = \varepsilon n_d + \alpha \frac{1}{2} n_d(n_d - 1)$$
$$+ \beta(n_d - \nu)(n_d + \nu + 3) + \gamma[L(L+1) - 6n_d], \tag{2.25}$$

where ε, α, β and γ are parameters that can be determined by fitting the calculated spectra to the experimental data. The level structure of the U(5) limit looks like those of phonon, in which the degeneracies are produced for the 4_1^+, 2_2^+ and 0_2^+ states, as well as 6_1^+, 4_2^+, 3_1^+, 2_3^+ and 0_3^+ states, ... etc. Electromagnetic transition rates can be obtained analytically, too.

SU(3) Limit

The second group chain is as follows,

$$\text{(chain II)} \quad U(6) \supset SU(3) \supset O(3) \supset O(2). \tag{2.26}$$

The basis set of chain II is characterized by the quantum number λ, μ, K (one related to the z component of angular momentum in the intrinsic frame), L and M. (λ, μ) labels the representation of the SU(3) group. The eigenvalue is obtained analytically:

$$E(\lambda, \mu, L, M) = \alpha L(L+1) - \beta[\lambda^2 + \mu^2 + \lambda\mu + 3(\lambda + \mu)], \tag{2.27}$$

where α and β are fixed by the fit. The SU(3) dynamical symmetry produces the rotational bands of the axially symmetric deformed nuclei.

2.3 Interacting Boson Model

O(6) Limit

The last group chain is

$$\text{(chain III)} \quad U(6) \supset O(6) \supset O(5) \supset O(3) \supset O(2). \tag{2.28}$$

The quantum numbers are σ (one characterizing the irreducible representation of the O(6) group), τ (d-boson seniority similar to v in U(5) limit), ν_Δ (same as n_Δ in U(5) limit), L and M. The eigenvalues are given by

$$E(\sigma, \tau, \nu_\Delta, L, M) = A\frac{1}{4}(N-\sigma)(N+\sigma+4) + B\frac{1}{6}\tau(\tau+1) + CL(L+1), \tag{2.29}$$

where A, B and C are adjusted to the data. It has been shown that the O(6) dynamical symmetry is associated with the γ unstable nuclei, in which the symmetry axis of the nucleus do not coincide with the intrinsic axis.

Nuclear collective spectra can be reproduced in a mathematically transparent way in terms of the dynamical symmetries of the IBM with remarkably small number of adjusted parameters. Nevertheless, vast majority of realistic nuclei do not exactly follow any of the three dynamical symmetries but are mix of these, and are therefore solved numerically in practice. The results presented in this thesis are obtained through the numerical diagonalization of the IBM Hamiltonian, rather than assuming any symmetry-dictated form of boson Hamiltonian. Note that, however, such algebraic feature of the IBM associated with a certain geometrical picture still holds and can serve as a paradigm for the description of collective structural evolution even when a realistic boson Hamiltonian is used.

2.3.2 Geometry

Since the IBM is the model for the quadrupole collective states, it is natural to think the model is associated with a certain geometrical picture. The geometrical aspect of IBM was discussed by Ginocchio and Kirson [33], by Dieperink et al. [34], and by Bohr and Mottelson [35], who introduced the so-called boson coherent state, i.e., the intrinsic wave function of the boson system. For the purpose of simplicity, we start with the IBM-1 description of the coherent state. The coherent state for the system composed of the monopole s and the quadrupole d bosons is defined as [33]

$$|\Phi(N, \{a_\mu\})\rangle = \frac{1}{\sqrt{N!}}\left(\lambda^\dagger\right)^N |0\rangle \quad \text{with} \quad \lambda^\dagger = \frac{1}{\sqrt{1+\sum_{\mu=-2}^{2}a_\mu^2}}\left(s^\dagger + \sum_{\mu=-2}^{2}a_\mu d_\mu^\dagger\right), \tag{2.30}$$

where $|0\rangle$ stands for the boson vacuum, e.g., inert core, and N represents the number of bosons. Here the set of the parameters $\{a_\mu\}$ in Eq. (2.30) is written more explicitly as

$$a_0 = \beta \cos\gamma, \quad a_{\pm 1} = 0, \quad \text{and} \quad a_{\pm 2} = \frac{1}{\sqrt{2}} \beta \sin\gamma, \tag{2.31}$$

where β and γ represent the intrinsic variables similar to the deformation parameters in the geometrical model [36]. The parameter β represents the relative d-boson probability amplitude over the s boson. As the s boson can create only a spherical state and the description of the quadrupole deformation requires the d boson, the βs are parameters indicating the quadrupole deformation. The coherent state $|\Phi\rangle$ represents an intrinsic state. If the quadrupole deformation has an axial symmetry, one can choose the z axis to be the symmetry axis. In this case, the coherent state must be invariant with respect to the rotation about the z axis. This leads us to $a_2 = a_{-2} = 0$, or $\gamma = 0°$ in Eq. (2.30). A different value of γ indicates a triaxial deformation. Thus, one can describe the (intrinsic) shape of the nucleus in terms of β and γ. The former measures the total magnitude of the deformation, while the latter the triaxiality. The energy surface for the system of interest is nothing but the expectation value of the corresponding boson Hamiltonian. The coherent state Eq. (2.30) can be then written in terms of the deformation variables β and γ

$$|\Phi(N,\beta,\gamma)\rangle = \{N!(1+\beta^2)\}^{-N/2} \left\{ s^\dagger + \beta\cos\gamma d_0^\dagger + \frac{1}{\sqrt{2}}\beta\sin\gamma(d_{+2}^\dagger + d_{-2}^\dagger) \right\}^N |0\rangle. \tag{2.32}$$

The coherent state for the IBM-2 system is given by [33]

$$|\Phi(N_\pi, N_\nu, a_{\pi,\mu}, a_{\nu,\mu})\rangle = \prod_{\rho=\pi,\nu} \frac{1}{\sqrt{N_\rho!}} \left(\lambda_\rho^\dagger\right)^{N_\rho} |0\rangle \tag{2.33}$$

with

$$\lambda_\rho^\dagger = \frac{1}{\sqrt{1+\beta_\rho^2}} \left\{ s_\rho^\dagger + \beta_\rho \cos\gamma_\rho d_{\rho,0}^\dagger + \frac{1}{\sqrt{2}}\beta_\rho \sin\gamma_\rho (d_{\rho,+2}^\dagger + d_{\rho,-2}^\dagger) \right\}. \tag{2.34}$$

In principle, both β_ρ and γ_ρ in Eq. (2.34) can take different values for proton and neutron bosons. Since protons and neutrons attract each other strongly, the proton and the neutron systems should have the same shape in the first approximation. We therefore assume that β_π (γ_π) and β_ν (γ_ν) take the same values, denoted by β_B (γ_B),[5] for proton and neutron bosons,

$$\beta_\pi = \beta_\nu \equiv \beta_B \quad \text{and} \quad \gamma_\pi = \gamma_\nu \equiv \gamma_B. \tag{2.35}$$

The range of γ_B is set $0° \leq \gamma_B \leq 60°$, as for the polar deformation parameter γ, which is valid as the boson and geometrical γs have the same meaning. Under these assumptions, the coherent state used in this thesis is written as

---

[5] The subscript B in the β and γ variables represents boson.

2.3 Interacting Boson Model

$$|\Phi(\beta_B, \gamma_B, N_\pi, N_\nu)\rangle = (1+\beta_B^2)^{-(N_\pi+N_\nu)/2} \prod_{\rho=\pi,\nu} \frac{1}{\sqrt{N_\rho!}}$$
$$\times \left\{ s_\rho + \beta_B \cos\gamma_B d^\dagger_{\rho 0} + \frac{1}{\sqrt{2}} \sin\gamma_B (d^\dagger_{\rho+2} + d^\dagger_{\rho-2}) \right\}^{N_\rho} |0\rangle,$$
(2.36)

where N_ρ stands for the proton and the neutron boson numbers. In the coherent state, the boson number is not well-defined, and can be arbitrary. However, this thesis employs the usual counting rule in such a way that the boson number in the coherent state always represents half the number of valence nucleons. The boson Hamiltonian will be diagonalized in the same configuration space.

2.3.3 Hamiltonian

In many of the IBM studies the Hamiltonian contains up to two body not only for the sake of simplicity but to preserve the dynamical symmetry.[6] To keep the discussion as general as possible, we start from a general two-body Hamiltonian.

$$\hat{H}_{\text{IBM}} = \hat{H}_\pi + \hat{H}_\nu + \hat{H}_{\pi\nu} \qquad (2.37)$$

where the first and the second terms

$$\hat{H}_\rho = E_0^{(\rho)} + \sum_{\alpha\beta} \varepsilon^{(\rho)} b^\dagger_{\rho,\alpha} b_{\rho,\beta} + \sum_{\alpha\beta\gamma\delta} u^{(\rho)}_{\alpha\beta\gamma\delta} b^\dagger_{\rho,\alpha} b^\dagger_{\rho,\beta} b_{\rho,\gamma} b_{\rho,\delta} \qquad (2.38)$$

stand for the interaction between like bosons, with $\rho = \pi$ (proton) or ν (neutron). The third term $\hat{H}_{\pi\nu}$ on the RHS of Eq. (2.37) represents the interaction between proton and neutron systems and is written generally as

$$\hat{H}_{\pi\nu} = \sum_{\alpha\beta\gamma\delta} w_{\alpha\beta\gamma\delta} b^\dagger_{\pi\alpha} b_{\pi\beta} b^\dagger_{\nu\gamma} b_{\nu\delta} + \cdots. \qquad (2.39)$$

The forms of the Hamiltonians in Eqs. (2.38) and (2.39) conserve the boson number. Here $\alpha, \beta, \gamma, \delta$ run from 1 through 6. The boson operator $b^\dagger_\rho$ is a shorthand notation of both s and d bosons, and is written explicitly as

$$b_{\rho 1} = s_\rho, \quad b_{\rho 2} = d_{\rho+2}, \quad b_{\rho 3} = d_{\rho+1}, \quad b_{\rho 4} = d_{\rho 0}, \quad b_{\rho 5} = d_{\rho-1}, \quad b_{\rho 6} = d_{\rho-2}.$$
(2.40)

Here $E_0^{(\rho)}$ in Eq. (2.38) is constant for a given nucleus and will be taken into account in Chap. 7. Note that the constant $E_0^{(\rho)}$ does not change the excitation energies.

---

[6] Specific form of the three-body boson term can be included in describing some cases of non-axially symmetric nuclei. This will be discussed in Chap. 6.

It is apparent that the general boson Hamiltonian of Eq. (2.37) cannot be used as it is in realistic calculations because of the large number of parameters. However, since only a few terms of the general IBM-2 Hamiltonian are essential in most of the realistic cases, one can adopt rather simplified Hamiltonian. The best known example would be the Hamiltonian in the so-called consistent-Q formalism of Warner and Casten in the IBM-1 case [37].

The most basic and widely-used IBM-2 Hamiltonian with up to two-body terms is written similarly to the consistent-Q formalism Hamiltonian in the IBM-1[7]:

$$\hat{H}_{\text{IBM}} = \varepsilon_\pi \hat{n}_{d\pi} + \varepsilon_\nu \hat{n}_{d\nu} + \kappa \hat{Q}_\pi \cdot \hat{Q}_\nu. \tag{2.41}$$

The first term on the RHS of Eq. (2.41) stands for the d-boson number operator,

$$\varepsilon_\rho \hat{n}_{d\rho} = \varepsilon d_\rho^\dagger \cdot \tilde{d}_\rho \tag{2.42}$$

with ε_ρ denoting the single d-boson energy relative to the s boson one. The proton ε_π and the neutron ε_ν parameters are usually set equal with each other since it is only their average that makes sense in actual nuclei. Thus the condition

$$\varepsilon_\pi = \varepsilon_\nu = \varepsilon \tag{2.43}$$

is assumed throughout in this thesis. The d-boson number operator with $\varepsilon > 0$ contributes to keeping a nucleus spherical. $d_\rho^\dagger \cdot \tilde{d}_\rho$ is a scalar product, and $\tilde{d}_\rho$ is defined as $\tilde{d}_{\rho\mu} \equiv (-1)^\mu d_{\rho-\mu}$ ($\mu = 0, \pm 1, \pm 2$). While ε can differ between proton and neutron, they are set equal for simplicity. The third term on the RHS of Eq. (2.41) is the quadrupole-quadrupole interaction between proton and neutron bosons with the strength κ, inducing the quadrupole deformation. The parameters χ_π and χ_ν appear as

$$\hat{Q}_\rho = [s_\rho^\dagger \times \tilde{d}_\rho + d_\rho^\dagger \times \tilde{s}_\rho]^{(2)} + \chi_\rho [d_\rho^\dagger \times \tilde{d}_\rho]^{(2)} \tag{2.44}$$

with $\tilde{s}_\rho \equiv s_\rho$, and determine the prolate or oblate shape of deformation, reflecting the structure of collective nucleon pairs as well as the numbers of valence nucleons [31, 32, 38].

Almost all situations of the low-lying quadrupole collective states can be described basically by the Hamiltonian of Eq. (2.41). The Hamiltonian of Eq. (2.41) is apparently not a general Hamiltonian, but embodies essential ingredients of the quadrupole collectivity. Moreover, it is such form of the Hamiltonian as in Eq. (2.41) that determines the basic topology of the energy surface. Let us consider the case of the quadrupole-quadrupole interaction as an example. Since the proton-neutron interac-

---

[7] Because of its proton-neutron two-fluid character, the IBM-2 Hamiltonian often contains the interaction term relevant to the symmetry energy (so-called Majorana term). However, since thorough assessment of the Majorana term, including its physical meaning, still remains to be done, we do not try to touch on this point.

2.3 Interacting Boson Model

tion is much dominant over the proton-proton and neutron-neutron interactions for medium-heavy and heavy deformed nuclei, the quadrupole-quadrupole interaction between like bosons, which drives the nucleus rather spherical, can be of less importance. As will be shown later, in some cases the Hamiltonian of Eq. (2.41) may not be good enough. In such cases some interaction terms should be taken into account.

Turning now to the intrinsic boson state, the expectation value of an operator $\hat{O}$ with respect to the coherent state $|\Phi(\beta_B, \gamma_B, N_\pi, N_\nu)\rangle$ in Eq. (2.36) is hereafter denoted by

$$\langle \hat{O} \rangle = \langle \Phi(\beta_B, \gamma_B, N_\pi, N_\nu) | \hat{O} | \Phi(\beta_B, \gamma_B, N_\pi, N_\nu) \rangle \quad (2.45)$$

unless otherwise specified. To calculate the bosonic energy surface (referred to as IBM energy surface, hereafter), IBM-2 Hamiltonian of Eq. (2.41) is substituted to $\hat{O}$. The IBM energy surface can be calculated straightforwardly, using the technique of Ref. [39], to have the analytical form as

$$\langle \hat{H}_{\text{IBM}} \rangle = \frac{\varepsilon (N_\pi + N_\nu)\beta_B^2}{1 + \beta_B^2} + N_\pi N_\nu \kappa \frac{\beta_B^2}{(1 + \beta_B^2)^2}$$
$$\times \left[4 - 2\sqrt{\frac{2}{7}}(\chi_\pi + \chi_\nu)\beta_B \cos 3\gamma_B + \frac{2}{7}\chi_\pi \chi_\nu \beta_B^2 \right], \quad (2.46)$$

which, depending on the values of the parameters ε, κ, χ_π and χ_ν, covers all geometrical pictures associated with three dynamical symmetries of the IBM: spherical vibrational, axially symmetric deformed (both prolate and oblate), and γ-unstable shapes, and the transitional shapes among these. Note that, however, neither of the coexisting and the triaxial minima can be described by the IBM energy surface in the case of Eq. (2.46), which is composed of one- and two-body boson interactions.

2.3.4 Other Boson Models

Besides the IBM-1 and the IBM-2 models, several extended versions of IBM have been developed. Although these extended IBM models are obviously out of the focus of this thesis, it should be of certain interest to have a brief look at these models since they have their own unique capabilities and open problems.

A straightforward extension of the model concerns, for instance, the possibility of including other types of bosons with angular momenta J^π other than 0^+ and 2^+, such as g ($J^\pi = 4^+$) and p ($J^\pi = 1^-$) bosons. The former has been often implemented into the sd space, either explicitly [40] or by renormalizing the fermion G pairs into SD space of fermions [41]. While, similarly to the original IBM, the simple algebraic relation holds even by the addition of g bosons, the actual numerical calculation becomes complicated. The importance of the g boson (or collective G pair) for the description of deformed nuclei will be discussed in Chap. 3. By introducing the spin 1^- boson, application of IBM to higher-lying giant dipole resonance has been considered by Rowe and Iachello [42].

Since in medium to heavy nuclei protons and neutrons occupy different major shells, the effect of the proton-neutron pair can be neglected in the standard sd-IBM, which assumes the proton-proton and the neutron-neutron pairs in the isospin $T = 1$ states only. However, for lighter nuclei particularly those with $N \approx Z$, in which protons and neutrons may occupy the same orbits, proton-neutron pair may become important and one should construct the boson Hamiltonian that holds isospin invariance. To reflect such shell structure of light nuclei, Elliott and White introduced the proton-neutron pair in $T = 1$ state (called IBM-3 [43]), and later proton-neutron pair in isospin $T = 0$ state was added by Elliott and Evans [44].

Other extensions of IBM worthy of noting would be the inclusion of unpaired, fermionic (single-particle) degrees of freedom in order to treat the odd-mass[8] and doubly-odd[9] nuclear structure (for review, see Ref. [45]). The first version of this model containing a single nucleon degree of freedom is called interacting boson-fermion model (IBFM),[10] which was introduced by Iachello and Scholten [46] and was developed so as to be feasible in realistic calculations. An important consequence of the boson-fermion algebras has been the peculiar pattern of level structures having the so-called dynamical supersymmetry [48], whose candidates have been found experimentally in a few limited cases of heavy nuclei (cf Ref. [45], and references therein). The concepts of IBFM and its supersymmeric feature have been further applied to the general theory of quantum phase transition in the system of bose-fermi mixture [49]. The boson-fermion models have been exploited mainly for heavy-mass nuclei, where even the underlying microscopic fermion structure are still not simple. Hence, further investigations will be necessary to establish a firm microscopic basis of these models. Also, even the simplest IBFM study considers only a single orbit, and hence this model should be generalized as to involve multi orbits. Much evidence for a set of supersymmetric nuclei should be investigated as well.

2.4 Nucleon-to-Boson Mapping

We are back now to the IBM-2 framework, and describe in what follows how the mapping from EDF to IBM can be done as well as the resultant spectra and transition strengths.

---

[8] Proton and neutron numbers are odd and even, respectively, or vice versa.

[9] Both proton and neutron numbers are odd.

[10] Distinction is not made between protons and neutrons. When another nucleon degree of freedom needs to be introduced in doubly odd nuclei, the IBFM is extended, which is called interacting boson-fermion-fermion model (IBFFM) developed e.g., by Brant et al. [47]. This version is of course quite convoluted.

2.4.1 Optimal Boson Hamiltonian

While the deformation variables β and γ could be in general different between fermion and boson systems, they can be related to each other, if the separability of the mapping along the β and the γ directions is assumed. It was shown by Ginocchio and Kirson [33] and by Dieperink et al. [34] that, in general, the bosonic and the geometrical β variables are proportional to each other and that the proportionality coefficient coincides with the ratio of the total nucleon number to the valence nucleon number counted from the nearest closed shells. We exploit this relation and assume that

$$\beta_B = C_\beta \beta_F \quad \text{and} \quad \gamma_B = \gamma_F, \tag{2.47}$$

where β_F and γ_F denote the geometrical deformation parameters of Eq. (2.18) and C_β the overall scaling factor characterizing the proportionality relation between fermion and boson β variables. The typical range of the C_β value turns out to be approximately 5–10, which is about the same order of magnitude as the actual ratios of the total nucleon number to the valence nucleon number. Note that C_β may vary from nucleus to nucleus gradually as a function of N_π and N_ν. Regarding the triaxial parameter γ, the equality $\gamma_B = \gamma$ seems valid as indeed both geometrical and IBM γ variables have the same meaning, ranging from 0 to 60°.

Alternatively, one can interpret the above-mentioned relationship between boson and geometrical β's like this: Under the assumption of axial symmetry, namely $\gamma_B = 0°$, the intrinsic quadrupole moment for boson system, denoted as Q_I, can be defined by the expectation value $Q_I = q \langle \hat{Q}_\pi + \hat{Q}_\nu \rangle$ with q being an overall scaling factor. Q_I is calculated as

$$Q_I = \frac{q\left[2(N_\pi + N_\nu)\beta_B - \sqrt{\frac{2}{7}}(N_\pi \chi_\pi + N_\nu \chi_\nu)\beta_B^2\right]}{1 + \beta_B^2}. \tag{2.48}$$

In the present study, the typical range of β_B is $0 \leq \beta_B \leq 1$. For $\beta_B \geq \sqrt{2}$, the system is too deformed, which is irrelevant to the present work. Also practically, $|\chi_\pi| \leq 1$ and $|\chi_\nu| \leq 1$ are fulfilled based on experience from the IBM phenomenology. Roughly speaking, the terms proportional to β_B^2 on the RHS of Eq. (2.48) then become minor as compared to the rest, and are neglected in the first good approximation. This leads us to $\beta_B \propto \beta_F$, because Q_I is associated with the mass quadrupole moment for fermion q_0 in Eq. (2.18) when the axial symmetry is assumed. Such rough estimation of β variables through Eq. (2.48) turns out to be consistent with the assumptions of Refs. [33, 34].

Once the $\beta\gamma$ deformation energy surface is obtained from the constrained self-consistent mean-field calculation with a fixed EDF, each point on the mean-field energy surface is mapped onto the corresponding point on the appropriate energy surface of boson system, as illustrated in Fig. 2.2. The procedure is exactly a mapping

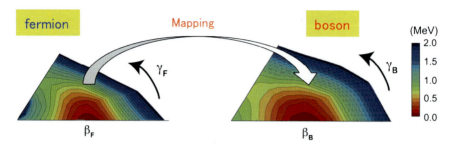

Fig. 2.2 Pictorial illustration of the fermion-to-boson mapping process. Each point on the constrained energy surface, obtained from the microscopic self-consistent mean-field calculation (*left-hand side* (LHS) of the figure), is mapped onto the appropriate point on the bosonic energy surface, which is obtained as the energy expectation value (RHS of the figure). Both energy surfaces are depicted in terms of the quadrupole collective coordinates up to 2 MeV excitation from the energy minima. While β_F and γ_F represent the geometrical (fermionic) deformation parameters β_B and γ_B their boson images in the IBM system. For details see the main text

of the mean-field state at the coordinate (β_F, γ_F) on the energy surface onto the corresponding IBM state at (β_B, γ_B). We actually determine the values of the five parameters, ε, κ, χ_π, χ_ν and C_β, for each individual nucleus by drawing the IBM energy surface with the formulae of Eqs. (2.46) and (2.47) so as to be identical to the self-consistent mean-field energy surface.

Since this formulation may sound somewhat too abrupt, we need to give a little more detailed account here. The basic topology of the self-consistent mean-field energy surface, which is, more specifically, the one around the absolute minimum, reflects essential fermion many-body properties including how the nuclear force and the Pauli principle work for determining the low-lying collective structure of the relevant intrinsic deformation. By matching the IBM Hamiltonian to the self-consistent mean-field energy surface with a fixed microscopic energy density functional, such basic properties of nuclear many-body system could be incorporated in a mathematically simpler, boson model.

To such an end, first of all a global pattern of the self-consistent mean-field energy surface should be reproduced, rather than its every detail. The overall global pattern of the microscopic energy surface in this context means, for instance, the location $(\beta_F, \gamma_F) = (\beta_{\min}, \gamma_{\min})$ on the $\beta\gamma$ surface at which the minimum occurs, and the curvatures with respect to β and γ, but not the local ones at a particular point on the energy surface. In determining the IBM parameters, special attention has to be paid also to their systematic changes with valence nucleon numbers, in accordance with the overall systematic change in the pattern of microscopic energy surface.

The scale factor C_β is mainly determined by adjusting the parameter values ($\beta_{\min}$, $\gamma_{\min}$), which give minimum, and the curvature in β direction, of the IBM energy surface to those of the original mean-field energy surface. The ε and κ parameters are mainly determined as functions of the boson numbers N_π and N_ν so as to reproduce the depth of the potential well in β direction (deformation energy), which is characterized by the energy difference between the configurations $\beta_F = \beta_{\min}$ and $\beta_F = 0$

2.4 Nucleon-to-Boson Mapping

with $\gamma = \gamma_{\min}$. As the γ dependence of the IBM energy surface of Eq. (2.46) appears as the term proportional to $\cos 3\gamma_B$, the softness of the mean-field energy surface in γ variable can be reproduced by adjusting the value of the quantity $\chi_\pi + \chi_\nu$, which also determines, depending on its sign, either a prolate or oblate minimum of the IBM energy surface. The parameters χ_π and χ_ν significantly depend on N_π and N_ν, respectively, but much less for vice versa. To reflect the structure of collective nucleon pairs, the sign of the χ_ρ parameter should be negative when the last proton/neutron occupy from the beginning to the middle of the major shell, i.e., the boson reflects pair of valence particles, while χ_ρ should be positive when the last proton/neutron surpass the middle of the major shell, i.e., the boson reflects pair of valence holes [32].

From a practical point of view, the χ-square fit may be a straightforward way to fix these parameters, but does not make much sense because of the local pattern of the microscopic energy surface. These local pattern cannot be reproduced by the IBM Hamiltonian of Eq. (2.41) because it is too simple for the complete fit. Nevertheless, the technique to avoid the problem has been introduced [59], and will be discussed in Sect. 2.4.2.

With the optimal set of the parameters thus obtained, we diagonalize the boson Hamiltonian of Eq. (2.41), to yield the levels and the wave functions having good quantum numbers in the laboratory system. The code NPBOS [50] has been a popular computer program open to the public, which diagonalizes the IBM-2 Hamiltonian in a set of bases coupled with total angular momentum (J-scheme). More recently the author developed a code which diagonalizes the IBM-2 Hamiltonian containing up to three-body interactions in a set of bases with $J_z = M$ (M-scheme).[11] In addition, as we will show in Chap. 7, eigenenergies of the IBM Hamiltonian formulated by the microscopic EDF may include to a good extent the quantum-mechanical effect that is missing in the static mean-field approximation.

Here we point out once again that a more general boson Hamiltonian with up to two-body terms may contain many other interaction terms. Some of these terms may affect excitation energies to a certain extent but the parameters of these terms cannot be determined by simply studying the self-consistent mean-field energy surfaces, because these terms vanish in the classical limit with assumptions of Eq. (2.35) and hence do not have any contribution to the energy surface. To fix strengths parameters of these interaction terms, one should go beyond the energy-surface mapping procedure and, in addition to this, should try to map some other quantities, e.g., energy shift of the ground state against infinitesimal rotation of it. A mapping scheme of this kind may become necessary for specific cases such as strongly deformed nuclei, which is beyond the scope of this chapter but will be investigated in detail in Chap. 3. In any case, while the form of the Hamiltonian $\hat{H}_{\text{IBM}}$ in Eq. (2.41) is simple, it embodies many aspects of quadrupole collective states.

What should be also noted is that, in some microscopic approaches of the collective model, the self-consistent mean-field energy surface is treated as an effective potential with subtraction of zero-point energies, and that the mass parameters of both rotational and vibrational motions are introduced to construct collective Hamil-

---

[11] The M-scheme diagonalization is described in Appendix B.2.

tonian. We point out that in the present work the total energy of the HF+BCS is compared with the corresponding energy of the IBM. In this case, the mass parameter itself does not show up explicitly, while its effect is, mainly for the systems with relatively weak quadrupole deformation, supposed to be included to a good extent in the diagonalization of the boson Hamiltonian calibrated by the comparison of energy surfaces.

Here the term "mapping", which is frequently used in this thesis, needs to be clarified further, since the term conveys slightly different meaning here from in the conventional fermion-boson mapping technique. In the conventional boson-mapping approach, the fermion angular momentum operator is represented by the boson creation and annihilation operators so that the basic algebra obeying the commutation relation of the group SU(2) should hold in boson system. These studies have confronted by their own problems of not having hermicity and/or not conserving the boson number, although possible solutions have been considered. The concept of boson mapping idea has been applied to the many-body problems like in the treatments of Belyaev-Zelevinski type [51] and of Marumori type [52]. The former study proposed to map the operators so that the commutation relations of all the physically relevant operators are preserved. In the latter the fermion states are mapped onto the corresponding boson states so that the matrix elements conserve. For the IBM, the treatment proposed by Otsuka et al. (OAI method) [32] should be similar to the boson mapping of Marumori et al., because in the OAI method the boson parameters are determined so that the matrix element of a fermion operator in the SD space is equated to that of the boson operator in the sd boson space. In the present work, the energy expectation value for fermion system is compared with the classical limit of the corresponding boson Hamiltonian in the boson coherent (intrinsic) state. Through this procedure, the mean effect of the fermion properties, not the operator itself, can be *mapped* onto the boson system. In this respect, while special attention should be paid to the use of the term "mapping", we will mean by the term the comparison of the energy expectation values for boson and fermion systems.

Energy Surfaces

In what follows actual mapping procedure is demonstrated by taking samarium ($_{62}$Sm) isotopes as an example. Figure 2.3 displays the constrained self-consistent HF+BCS and the IBM energy surfaces for $^{146-156}$Sm isotopes. Two different but widely reputed parameterizations SLy4 and SkM* are drawn on the left-hand side (LHS) and the RHS of Fig. 2.3, respectively, in order to show the generality of the procedure. The energy surfaces are depicted in contour plots within the excitation of 2 MeV in energy measured from the potential minimum. The coordinate β_B (γ_B) of the IBM energy surface is expressed in terms of β_F (γ_F), for simplicity, using the formula of Eq. (2.47). We show the energy surfaces up to 2 MeV excitation since the low-lying collective states are supposed to be dominant in this range. We also note that the deformation parameter β (γ) that appears in the following always means β_F (γ_F), unless otherwise specified.

2.4 Nucleon-to-Boson Mapping

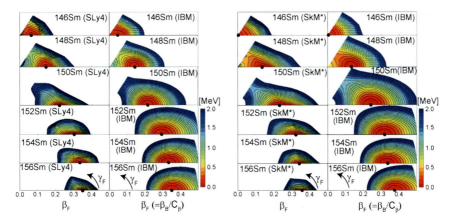

Fig. 2.3 The deformation energy surfaces in $\beta\gamma$ plane, obtained from the self-consistent constrained HF + BCS mean-field calculation with Skyrme SLy4 (*left column*) and SkM* (*right column*) functionals, and those of the IBM. These energy landscapes are drawn up to 2 MeV in energy from the absolute minima. The coordinate β_B (γ_B) of the IBM energy surface is expressed in terms of β_F (γ_F) using the formula of Eq. (2.47). γ_F is limited to 0–60°. Contour spacing is 100 keV and minima can be identified by the *solid circles*. The figures have been taken from Ref. [59]

Figures 2.3 shows that both Skyrme SkM* and SLy4 functionals give similar energy surface, while there exist certain differences particularly in the transitional nuclei. One can find abrupt changes in HF+BCS energy surface from $N = 86$ to 88 for the SLy4 case and from $N = 88$ to 90 for the SkM*. For $N = 84$ and 86 nuclei, one also finds a difference between the SLy4 and SkM* functionals. Namely, the latter is a bit flatter in both β and γ directions than the former. With the increase of the neutron number, N, the HF+BCS energy surface becomes steeper in both β and γ directions and $\beta_{\min}$ shifts away from the origin, resulting in well isolated prolate minimum for larger N. The nucleus ^{148}Sm has been recognized as an example of the spherical vibrator, being close to U(5) limit of IBM. However, the Skyrme HF+BCS energy surface in the present calculation for ^{148}Sm somewhat differs from this picture, placing the energy minimum at $\beta \sim 0.15$. The IBM energy surface reproduces $\beta_{\min}$ and the overall pattern of the HF+BCS energy surface. In the vicinity of SU(3) limit, ^{154}Sm, HF+BCS energy surface has a pronounced sharp minimum, and IBM energy surface also exhibits a similar one. The minimum valley is, however, shallower for the IBM energy surface. This is a general trend that cannot be changed any more by simply playing with parameters, and is probably due to the finiteness and/or limitation of boson configuration space.

Parameters of Boson Hamiltonian

In Fig. 2.4 shown are the derived parameters of the IBM Hamiltonian, ε, κ, χ_π, χ_ν and C_β, as functions of the neutron number N. The parameters in Fig. 2.4a–f were fixed

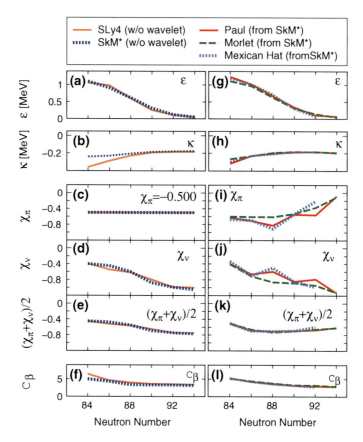

Fig. 2.4 a–f Evolution of the IBM parameters for Sm isotopes as functions of N. The wavelet analysis has not been performed. The Skyrme SLy4 (*solid curve*) and SkM* (*dotted curve*) functionals are used. The fixed value $\chi_\pi = -0.500$ is used for both Skyrme forces. **g–l** Evolution of the optimized parameters by the wavelet analysis, which are discussed in Sect. 2.4.2. The Paul (*solid curve*), Morlet (*dashed curve*) and Mexican Hat (*dotted curve*) wavelet functions are used. The figure has been taken from Ref. [59]

by the best fit to the microscopic self-consistent mean-field energy surface. There is no strict rule for this, and one should more or less refer to the experience from the IBM phenomenology and existing microscopic IBM studies[12] in determining these parameter values. The derived parameters in Fig. 2.4a–f are therefore not unique, and may be ambiguous. To avoid the ambiguity, we have formulated an automated way [59], using the wavelet transform [60], that finds the unique set of IBM parameters for a given nucleus. In this section, however, we would rather sketch the basic outline of the mapping procedure and how it works. Therefore we first discuss here the

---

[12] For instance, the parameter $\chi_\rho < 0$ takes the value within the range $|\chi_\rho| \leq 1$. The SU(3) limit of the parameter χ_ρ is $\sqrt{7}/2$. The value much larger than $\sqrt{7}/2$ does not make any physical sense.

2.4 Nucleon-to-Boson Mapping

parameters in Fig. 2.4a–f, and will introduce the unambiguous procedure in the next section (Sect. 2.4.2). In addition, at a number of places the parameters derived by the present work are compared with the earlier studies of IBM phenomenology and of the Otsuka-Arima-Iachello (OAI) method using shell-model interaction [31, 32].

In general the parameters ε and χ_ν in Fig. 2.4a, d, respectively, vary rather significantly, while κ and C_β change much less. In Fig. 2.4a the parameter ε becomes smaller with the neutron number N similarly to the earlier phenomenological study [53]. Why the gradual decrease of ε with N occurs remains to be clarified, but has been discussed in a microscopic picture [54] as a consequence of stronger coupling between "unperturbed d boson" and other types of bosons such as the one with spin 4 [or g boson] [31, 32, 40, 41]. In Fig. 2.4b, the magnitude of κ is in general set somewhat large in comparison to the phenomenological value [53] and to the OAI result [55]. This indicates that the present HF+BCS energy surface for Sm isotopes exhibits a too deep potential valley to be reproduced by the IBM energy surface with the phenomenological value of κ. For $N = 84$ and 86 nuclei, the κ value for the SLy4 EDF deviates significantly from that of the SkM* one, reflecting the quantitative difference of the microscopic energy surface between the SLy4 and the SkM* EDFs in Fig. 2.3. As seen from Fig. 2.4c, a fixed value of $\chi_\pi (= -0.500)$ is assumed for simplicity in both SLy4 and SkM* results, similarly to the results of the OAI method [31, 32, 38]. In Fig. 2.4d, the parameter χ_ν has a rather strong dependence on the neutron number N and changes at $N = 88$ or 90, and thereby reflects the structural evolution from transitional to deformed shapes. The seniority prescription in the OAI mapping [31, 32] gives the opposite dependence of χ_ν on N, while the present one appears to be consistent with a mapping method using deformed intrinsic states [54]. The average $(\chi_\pi + \chi_\nu)/2$ is also shown in Fig. 2.4e, which will be discussed in Sect. 2.4.2. The scale factor C_β in Fig. 2.4f becomes smaller in smooth systematics with N, reflecting the gradual shift of the quadrupole deformation identified by $\beta_{\min}$. In Fig. 2.4f, for deformed nuclei with $N \geq 92$, the value of C_β satisfies that $C_\beta \beta_{\min}$ is smaller than $\sqrt{2}$, at which the minimum occurs in SU(3) limit with infinite boson number $N_\rho \to \infty$ [33, 35, 39].

Excitation Energies

In Fig. 2.5 we show the low-lying spectra for Sm isotopes with $N = 84$–94 as functions of the neutron number N. The experimental spectra are shown in Fig. 2.5a, in which each data is connected by a line to guide eyes. Figure 2.5b, c show the theoretical level energies resulting from SLy4 and SkM* functionals, respectively, using the derived parameters in Fig. 2.4a–f. Figure 2.5d–f show the results using the IBM parameters determined by the wavelet analysis, which will be discussed in Sect. 2.4.2.

At both $N = 84$ and 86 nuclei in Fig. 2.5b, c, the calculated spectra exhibit U(5)-like features: 4_1^+, 2_2^+ and 0_2^+ states form a triplet. However, this is not the case with the experimental data for $N = 84$, in which 6_1^+ state is lying close to 4_1^+,

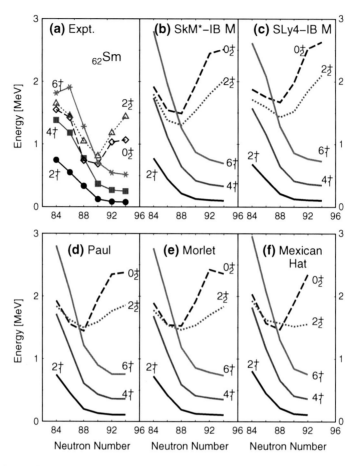

Fig. 2.5 Low-lying spectra for Sm isotopes as functions of N. *Upper panel*: **a** the experimental [61] and calculated excitation energies with Skyrme **b** SkM* and **c** SLy4 forces. *Lower panel*: Results from the wavelet analysis, which will be discussed in Sect. 2.4.2, using **d** Paul, **e** Morlet, and **f** Mexican Hat wavelets. The figure has been taken from Ref. [59]

2_2^+ and 0_2^+ states. This means that, for the ^{146}Sm nucleus, the 4_1^+ state may not be constructed in the sd boson model. For $N = 86$, the calculated levels are fairly close to the experimental data. With the increase of N, each calculated level comes down consistently with the experimental trends particularly for the yrast levels.

As noted empirically, there should be a critical point at $N = 90$, beyond which the 2_2^+ and 0_2^+ levels turn to go up with N. In Fig. 2.5b, c the sharp increases of the calculated side-band levels occur at $N = 88$, rather than at $N = 90$. For the $N \geq 92$ nucleus, one sees rotational spectra. The OAI mapping for Sm isotopes [55], where the collective G pair is renormalized perturbatively, gives similar results, but the present calculation seems to better reproduce the trends of the side-band levels including their transitional behaviors.

2.4 Nucleon-to-Boson Mapping

The overall systematic trend of the level energies in Fig. 2.5a–c reveals properties of X(5) critical-point symmetry [56–58] around $N = 88$ or 90, while the $N = 88$ nucleus is closer to the X(5) model in the present work. In fact, the calculated values of the ratio $R_{4/2} \equiv E_x(4_1^+)/E_x(2_1^+)$ [13] for $N = 88$ and 90 for SLy4 (SkM*) functional are 3.09 (3.01) and 3.30 (3.30), respectively. The experimental $R_{4/2}$ value for the ^{152}Sm nucleus and the X(5) one are 3.01 and 2.91, respectively. This deviation may be due to the properties of the HF+BCS energy surfaces on the RHS panels of Fig. 2.3, which do not identify a particular nucleus as the critical point.

What is worth noting here is that the calculated excitation levels for $N \geq 90$ are higher than the experimental [61] and the phenomenological [53] ones, although the overall pattern characterized, e.g., by the ratios between levels, is reproduced fairly well with a clear signature of the spherical-to-axially-deformed shape phase transition. This problem seems to be seen in many GCM results [58, 62] for well deformed nuclei, but never arises in moderately deformed cases. The problem with the rotational band is closely linked to the microscopic justification of IBM for rotational deformed nuclei. This kind of argument is beyond the scope of this chapter, and will be presented more in detail in Chap. 3.

$B(E2)$ Ratios

Having calculated the wave functions of the excited states, one is subsequently able to obtain other spectroscopic observables using these wave functions, among which the reduced E2 transition strength $B(E2)$ between the states with angular momenta L and L' is of particular importance. Using the Wigner-Eckart theorem, the $B(E2)$ is defined as [36]

$$B(E2; L \to L') = \frac{1}{2L+1}|\langle L'||\hat{T}^{(E2)}||L\rangle|^2, \tag{2.49}$$

where $\langle L'||\hat{T}^{(E2)}||L\rangle$ denotes the reduced matrix element which does not depend on the $L_z = M$ quantum number. Here the E2 transition operator $\hat{T}^{(E2)}$ is given by [3, 4]

$$\hat{T}^{(E2)} = e_\pi \hat{Q}_\pi + e_\nu \hat{Q}_\nu, \tag{2.50}$$

where e_ρ represents the boson effective charge. To reduce the number of free parameters, the operator $\hat{Q}_\rho$, which appears on the RHS of Eq. (2.50), can be identified as the quadrupole operator in Eq. (2.44) [37]. The same values of χ_π and χ_ν parameters as those derived from energy surface can be used for calculating the $B(E2)$ values. In principle, the boson charge e_ρ could be determined by taking into account the effect e.g., of core polarization, which goes beyond the mean-field approximation. In the mapping procedure presented in this thesis, the boson effective charge is thus the only adjustable parameter.

---

[13] $E_x(4_1^+)$ and $E_x(2_1^+)$ denote the 4_1^+ and the 2_1^+ excitation energies, respectively.

In this chapter, we assume $e_\pi = e_\nu$, for simplicity, and focus our discussion on the $B(E2)$ ratios defined as follows.

$$R_1 = \frac{B(E2; 4_1^+ \to 2_1^+)}{B(E2; 2_1^+ \to 0_1^+)}$$

$$R_2 = \frac{B(E2; 2_2^+ \to 2_1^+)}{B(E2; 2_1^+ \to 0_1^+)}$$

$$R_3 = \frac{B(E2; 0_2^+ \to 2_1^+)}{B(E2; 2_1^+ \to 0_1^+)}$$

$$R_4 = \frac{B(E2; 2_2^+ \to 0_1^+)}{B(E2; 2_2^+ \to 2_1^+)}, \qquad (2.51)$$

which, as we shall show, exhibit the traces of the shape-phase transitions as functions of the neutron number N.

Figure 2.6 shows the $B(E2)$ ratios R_1–R_4 for Sm isotopes as functions of N, studied with SLy4 and SkM* functionals. Experimental data [63–66] are shown as well. There is no significant difference between SLy4 and SkM* results.

The calculated R_1 value decreases with N and becomes close to the SU(3) limit of the IBM, $R_1 = \frac{10}{7}$ (indicated by dotted line), which is fairly consistent with the experimental data. The calculated values of R_2 and R_3 appear to indicate the

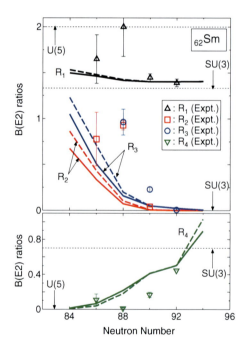

Fig. 2.6 $B(E2)$ ratios for Sm isotopes as functions of N. the ratios R_1–R_4 are defined in Eq. (2.51). *Solid* and *dashed curves* stand for the calculated results with SLy4 and SkM* forces, respectively. Experimental data are taken from Refs. [63–66]. The figure is taken from Ref. [59]

2.4 Nucleon-to-Boson Mapping

transition from spherical to deformed shapes, changing smoothly from around the U(5) limit (= 2) to the SU(3) limit (= $\frac{10}{7}$), while for the transitional $N = 88$ nucleus, the calculated values are smaller than both the experimental data and the phenomenological value [53].

The calculated R_4 increases as a function of N consistently with the experiments. In particular, it is quite close to U(5) limit, $R_4 = 0$, and SU(3) limit, $R_4 = \frac{10}{7}$ at $N = 86$ and 92, respectively. For $N = 88$ and 90 nuclei, similarly to the R_2 and the R_3 ratios, the present R_4 values suggest that these nuclei are already rotational, compared with the experimental data.

2.4.2 Uniqueness of the Boson Parameters

In Sect. 2.4.1, we noted that the physically relevant IBM parameters can be derived so that the IBM energy surface reproduces the microscopic mean-field energy surface as much as possible. The χ-square fit seems to be a straightforward way one may think of. However, it turns out that the χ-square fit does not work because of local patterns in the self-consistent microscopic energy surface. The χ-square fit provides many different combinations of the derived IBM parameters, which can give perfect fit but usually do not make any physical sense. It may be then questioned whether or not the physically relevant IBM parameters can be determined both unambiguously and uniquely.

In this respect, we introduce the wavelet analysis and show that, with the help of it, the global feature of the microscopic self-consistent mean-field energy surface, that is relevant to the low-lying collective states, can be extracted naturally. Originally, the wavelet technique has been developed in the field of signal processing (for reviews, see Refs. [60, 67], for instance) and also applied to a physical system like giant resonance phenomena [68]. In a general theory of the wavelet method, a given signal is transformed into a set of coefficients (wavelet transform) with respect to an appropriate basis function (a wavelet function or a so-called wavelet). The wavelet function is localized both in time and frequency domains. The wavelet function (denoted by Ψ) must have the properties that it has zero mean and that it is square integrable (so-called admissibility condition) [60]:

$$\int_{-\infty}^{\infty} \Psi^*(x) dx = 0 \quad \text{and} \quad K_\Psi \equiv \int_{-\infty}^{\infty} |\Psi(x)|^2 dx < \infty, \tag{2.52}$$

which mean that Ψ must oscillate in a finite duration. These conditions allow one to analyze efficiently the localized signal, some part of which is particularly important like the relevant low-energy region around the absolute minimum of the microscopic constrained energy surface. In addition, one is able to choose a wavelet which appears to be suited well for extracting a characteristic feature of the signal in question. These flexibilities make the wavelet analysis distinct from the Fourier transform, which localizes only frequency with limitation in use of a basis function. When applied to the present case, the wavelet transform of the energy surface can be characterized by the deformation variable β (γ) and its scale parameter $\delta\beta$ ($\delta\gamma$), which, in the language of the signal processing, correspond respectively to time and frequency.

To perform the analysis as precisely as possible, the continuous wavelet transform should be carried out, rather than the discrete transform [68]. The wavelet transform of the energy surface in β direction (for fixed γ) is formulated as

$$\tilde{E}(\delta\beta, \beta) = \frac{1}{\sqrt{\delta\beta}} \int E(\beta', \gamma) \Psi^* \left(\frac{\beta - \beta'}{\delta\beta} \right) d\beta', \quad (2.53)$$

where Ψ^* is a complex conjugate of a wavelet function Ψ. $E(\beta', \gamma)$ stands for the energy surface of either the HF + BCS or the IBM, while $\tilde{E}(\delta\beta, \beta)$ is its wavelet transform and is in general a complex value. Note that the wavelet transform is done separately for β and γ directions because the integral in Eq. (2.53) is defined in one dimension. Therefore, for γ direction, one has only to replace β ($\delta\beta$) with γ ($\delta\gamma$) in Eq. (2.53).

One should in general choose a wavelet function Ψ having a somewhat similar shape as the original signal. Then, we employ the following wavelet functions, which are frequently used [60, 67]

- Mexican Hat: $\Psi(x) = (1 - x^2) \exp\left(-\frac{x^2}{2}\right)$
- Morlet: $\Psi(x) = \pi^{-1/4} e^{ikx} \cdot e^{-x^2/2}$
- Paul: $\Psi(x) = 2^m i^m m! (1 - ix)^{-(m+1)} / \sqrt{\pi(2m)!}$,

where integers k (in Morlet) and m (in Paul) are control parameters and are set as $k = 6$ and $m = 4$, respectively, for practical reasons. Mexican Hat is nothing but the second derivative of the Gaussian function. These wavelets exhibit in common an oscillation within a Gaussian-like envelope.

The integration in Eq. (2.53) is performed by means of the fast Fourier transform [67]. Here special attention has to be paid for choosing appropriate integration range of β' (or γ'), i.e., the relevant range of low-energy excitation (typically up to several MeV from the absolute minimum). For strongly deformed nuclei, in particular, the integration range in β direction should be as small as possible and should include $\beta_{\min}$. It is only with such choice of the integration range that one is able to obtain physically relevant parameter set. Note that any region in the microscopic energy surface with β much larger than $\beta_{\min}$ does not have to be reproduced, because the topology of the energy surface for large β value is determined mainly by non-collective (single-nucleon) degrees of freedom.

In addition, one is able to reconstruct the original signal out of $\tilde{E}(\delta\beta, \beta)$ in Eq. (2.53). By doing so, one may ensure the wavelet transform is done properly. For fixed γ, the reconstructed energy surface is written as

$$E(\beta, \gamma) = \frac{1}{K_\Psi} \int \int \frac{\tilde{E}(\delta\beta, \beta')}{(\delta\beta)^{5/2}} \Psi \left(\frac{\beta' - \beta}{\delta\beta} \right) d(\delta\beta) d\beta'. \quad (2.54)$$

2.4 Nucleon-to-Boson Mapping

Actually, we first calculate the squared wavelet transform of the IBM energy surface and fit it to that of the constrained HF+BCS energy surface. Then, we optimize the IBM parameters by the simplex method. Thereby the IBM parameters can be derived without any arbitrariness. The γ degrees of freedom are also taken into account by minimizing the sum of the χ-square functions for β and γ directions. We use the mesh spacings $\Delta\beta = 0.01$ and $\Delta\gamma = 10°$. In the following, the ^{148}Sm and the ^{152}Sm nuclei are taken as examples, studied with SkM* functional. The Paul wavelet is used. The wavelet transforms are performed for $\gamma = 0°$ in the ranges $-0.29 \leq \beta \leq 0.29$ for ^{148}Sm and $-0.24 \leq \beta \leq 0.38$ for ^{152}Sm. Concerning the triaxial degrees of freedom, the minima on the lines of $\gamma = 10°, 20°, 30°, 40°$ and $50°$, in addition to $\gamma = 0°$ and $60°$, are also taken for the fits.

In each panel of Fig. 2.7, the squared amplitude $|\tilde{E}|^2$ is drawn in the contour plot for each scale value $\delta\beta$ on the vertical axis and for the β value on the horizontal axis. The left panels of Fig. 2.7 show the wavelet transforms of the self-consistent HF+BCS energy surfaces (denoted by MF-SkM*), while the right panels show the corresponding transforms of the IBM energy surfaces. $|\tilde{E}|^2$ is remarkably large for both nuclei when $\delta\beta$ is of the order of 10^{-1}, where the characteristic features of the energy surface can be most clearly seen. For the position (β) domain, the location where the amplitude $|\tilde{E}|^2$ becomes largest does not necessarily correspond to the β_{min} of the energy surface. In fact, for ^{148}Sm (^{152}Sm), $\beta_{min} \sim 0.15$ (0.32) as one sees from the RHS panels in Fig. 2.3, whereas $|\tilde{E}|^2$ is relatively large within the range $-0.30 \leq \beta \leq -0.10$ ($0.10 \leq \beta \leq 0.30$) in Fig. 2.7. Nevertheless, the pattern of the amplitude $|\tilde{E}|^2$ seems to have a certain relevance to that of the original energy surface. The magnitude of the former at a point in the position domain appears to reflect the slope of the latter at the corresponding point. The pattern of $|\tilde{E}|^2$ is not strongly affected by the local feature of the energy surface. Therefore, the energy surface fit in the wavelet space can be, as expected, suitable for considering the characteristic, global feature of the energy surface over the range of interest. By the fit with the wavelet analysis outlined above, the patterns of $|\tilde{E}|^2$s are to every detail in good agreement between the HF+BCS and the IBM for both ^{148}Sm and ^{152}Sm nuclei. Other types of the wavelets give the different pattern of $|\tilde{E}|^2$s from the Paul case, but the same argument applies.

In Fig. 2.8a, c, both the original (dashed curves) and the reconstructed (solid curves) HF+BCS energy surfaces for ^{148}Sm and ^{152}Sm are depicted in real space. The reconstructed energy surface is drawn using the formula in Eq. (2.54) with $\gamma = 0°$ and agrees with the original one. This means the wavelet transform is done properly. The corresponding IBM energy surfaces are also shown in Fig. 2.8b, d and are compared with the HF+BCS energy surfaces. Note that the IBM energy surface is depicted so that its origin point agrees with the point of the HF+BCS mean-field energy surface at the spherical configuration, ($\beta_F = 0, \gamma_F = 0°$). One can see that the HF+BCS energy surface is reproduced nicely by the IBM.

We show on the RHS panels of Fig. 2.4 the evolution of the IBM parameters for Sm isotopes, derived by the wavelet analysis. The Paul, Morlet and Mexican Hat wavelets are used. The IBM parameters from SkM* functional, which appear on the LHS panels of Fig. 2.4, have been chosen as initial guesses for the simplex method

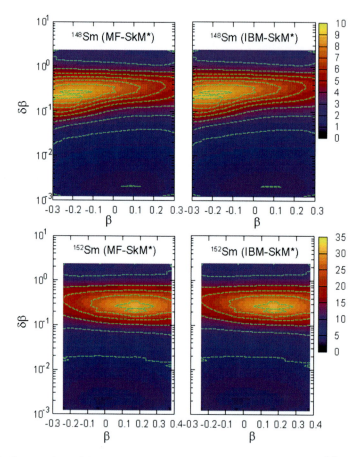

Fig. 2.7 Contour plots of the squared wavelet transforms, i.e., the amplitudes $|\tilde{E}|^2$, for the self-consistent constrained mean-field calculation with the HF+BCS model (denoted by MF-SkM*) and for the IBM energy surfaces (denoted by IBM-SkM*) in the β–$\delta\beta$ planes for the ^{148,152}Sm nuclei. The Paul wavelet is used. Note that the vertical axis is in logarithmic scale. The figure is taken from Ref. [59]

and are hereafter referred to as "*w/o*-wavelet" parameters, as indicated in the upper panel of Fig. 2.4. The initial parameters are not chosen arbitrarily, but are more or less those close to the *w/o*-wavelet parameters. This can be tested by using the same *w/o*-wavelet parameters as initial guesses, one of which is, however, replaced by a different value far from the *w/o*-wavelet one. In this case, the χ-square fit does not give the global minimum. In addition there may be some other combinations of the IBM parameters which give good fits, but in most cases such parameters do not make much physical sense.

In the present wavelet analysis, we treat χ_π as a free parameter. Note that the parameters for the Mexican-Hat wavelet at the $N = 94$ nucleus are not shown since the Mexican-Hat wavelet does not work there due to some technical problem. In

2.4 Nucleon-to-Boson Mapping

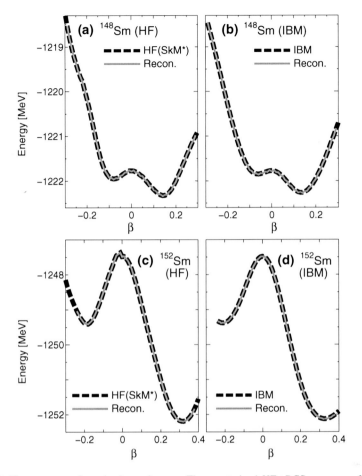

Fig. 2.8 The energy surfaces in the real space. The constrained HF+BCS energy surfaces for **a** ^{148}Sm and **c** ^{152}Sm are compared with the IBM energy surfaces for **b** ^{148}Sm and **d** ^{152}Sm, respectively. The reconstructed energy surfaces ("Recon.") are also plotted. The figure is taken from Ref. [59]

Fig. 2.4g–l, all three wavelet functions give almost identical values of ε, κ and C_β, while χ_π and χ_ν somewhat depend on the wavelets. For Paul and the Mexican-Hat cases, both χ_π and χ_ν do not evolve smoothly like that for the w/o-wavelet fits, whereas the Morlet wavelet seems to show fairly consistent trends of χ_π and χ_ν with w/o-wavelet results. As one sees from Fig. 2.4k, however, the average of χ_π and χ_ν for wavelet calculations has no notable dependence on the choice of the wavelets, while it is less sensitive to N compared to that of the w/o-wavelet calculation. In the future it may be necessary to make it clear why the difference of the local patterns of χ's occurs depending on the choice of the wavelets, and to conclude which wavelet is the best for a given physical system. As one sees from the comparisons in Fig. 2.4, it can be revealed that the optimized parameters by the wavelet analysis are almost the same as the *w/o*-wavelet ones, except for some local behaviors of the parameters. Some uncertainties of the optimized parameters obviously exist, and probably come from either the properties of the wavelet functions, or from other choice of the control parameters for numerical use, but they do not affect the resulting levels qualitatively.

The lower panel of Fig. 2.5 shows the evolution of the calculated excitation spectra with the parameters obtained from the wavelet analyses for (d) Paul, (e) Morlet and (f) Mexican Hat wavelets. Here we emphasize that any of these wavelet functions gives almost the same results to the *w/o*-wavelet calculation, which is consistent with the experimental trends including the transitional behaviors.

2.5 Brief Summary

In this chapter the method of deriving the Hamiltonian of the IBM based on the constrained self-consistent mean-field approach using microscopic nuclear EDF has been introduced. The fermionic constrained energy surface with quadrupole degrees of freedom was mapped onto an appropriate classical limit of the boson Hamiltonian in the coherent-state formalism. The strength parameters of the IBM Hamiltonian can be determined through this process. The IBM Hamiltonian parametrized by these interaction strengths is diagonalized, providing excitation energy and the electromagnetic transition rates. As a first stringent test, axially symmetric Sm isotopes has been analyzed starting from the constrained Skyrme HF+BCS method with the SkM* and SLy4 functionals, and the observed structural evolution from the vibrational to the rotational states has been reproduced well.

The results of the present work exhibited almost equal quality of agreement with experiment to the earlier OAI mapping that uses an empirical shell-model interaction. However, a shell-model interaction may not be always an appropriate starting point because it contains too many aspects. Some of these aspects are enhanced, but some others may be suppressed in the low-lying collective states. A popular energy density functional, any of Skyrme and Gogny as well as the RMF functionals, is universal in such a way that it is already calibrated to the observed bulk properties of finite nuclei, and is hence supposed to be more appropriate to start with for the analysis of nuclear shape phenomena. Although the Skyrme EDF has been used throughout this chapter, the procedure turns out be, as one will see later, general so that it is almost independent of the details of EDF.

It was shown that the IBM parameters can be derived unambiguously by using the technique of the Wavelet transform, whereas naive χ-square fit does not make sense. Wavelet transform is done first for the self-consistent mean-field energy surface within the appropriate range within the $\beta\gamma$ plane that is relevant to the low-lying quadrupole collective states, and the wavelet transform of the fermionic energy surface is subsequently fitted to the relevant Wavelet transform of the IBM energy surface. This procedure is quite robust because only essential feature of fermionic energy surface, rather than every detail, can be extracted quite effectively.

2.5 Brief Summary

The present mapping idea opens up a new avenue for calculating the collective quadrupole dynamics of medium-heavy and heavy nuclei from a microscopic picture, and a particularly important outcome is such that, contrary to earlier empirical IBM studies, we gain a capability to predict energies and wave functions of excited states in unknown territories on the nuclear chart, not yet studied experimentally.

References

1. Ring P, Schuck P (1980) The nuclear many-body problem. Springer, Berlin
2. Bender M, Heenen P-H, Reinhard P-G (2003) Self-consistent mean-field models for nuclear structure. Rev Mod Phys 75:121–180
3. Arima A, Iachello F (1975) Collective nuclear states as representations of a SU(6) group. Phys Rev Lett 35:1069
4. Iachello F, Arima A (1987) The interacting boson model. Cambridge University Press, Cambridge
5. Hohenberg P, Kohn W (1964) Inhomogeneous electron gas. Phys Rev 136:B864
6. Kohn W, Sham LJ (1965) Self-consistent equations including exchange and correlation effects. Phys Rev 140:A1133
7. Skyrme THR (1959) The effective nuclear potential. Nucl Phys 9:615
8. Vautherin D, Vén Veroni M (1969) A Hartree-Fock calculation of 208Pb in coordinate space. Phys Lett B 29:203
9. Vautherin D, Brink DM (1972) Hartree-Fock calculations with Skyrme's interaction. I. Spherical nuclei. Phys Rev C 5:626
10. Decharge J, Girod M, Gogny D (1975) Self consistent calculations and quadrupole moments of even Sm isotopes. Phys Lett B 55:361
11. Dechargé J, Gogny D (1980) Hartree-Fock-Bogolyubov calculations with the D1 effective interaction on spherical nuclei. Phys Rev C 21:1568
12. Walecka JD (1974) A theory of highly condensed matter. Ann Phys 83:491
13. Brink DM, Boeker E (1967) Effective interactions for Hartree-Fock calculations. Nucl Phys A 91:1
14. Gogny D (1975) Simple separable expansions for calculating matrix elements of two-body local interactions with harmonic oscillator functions. Nucl Phys A 237:399
15. Berger JF, Girod M, Gogny D (1984) Microscopic analysis of collective dynamics in low energy fission. Nucl Phys A 428:23c
16. Chappert F, Girod M, Hilaire S (2008) Towards a new Gogny force parametrization: impact of the neutron matter equation of state. Phys Lett B 668:420
17. Goriely S, Hilaire S, Girod M, Peru S (2009) First Gogny-Hartree-Fock-Bogoliubov nuclear mass model. Phys Rev Lett 102:242501
18. For instance, Abe D (2008) New type of density-dependent effective interaction and its applications to exotic nuclei. Ph.D. thesis, University of Tokyo.
19. Vretenar D, Afanasjev AV, Lalazissis GA, Ring P (2005) Relativistic HartreeBogoliubov theory: static and dynamic aspects of exotic nuclear structure. Phys Rep 409:101
20. Nikšić T, Vretenar D, Ring P (2011) Relativistic nuclear energy density functionals: mean-field and beyond. Prog Part Nucl Phys 66:519
21. Bonche P, Flocard H, Heenen PH (2005) Solution of the Skyrme HF + BCS equation on a 3D mesh. Comput Phys Commun 171:49
22. Bertsch GF (2005) Nuclear structure in mean-field theory and its extensions. Lecture note at the 2005 summer school of the Center for Nuclear Study (CNS), University of Tokyo (at the Wako campus of RIKEN).

23. Schunck N, Dobaczewski J, McDonnell J, Satuła W, Sheikh JA, Staszczak A, Stoitsov M, Toivanen P (2012) Solution of the Skyrme-Hartree-Fock-Bogolyubov equations in the Cartesian deformed harmonic-oscillator basis. (VII) hfodd (v2.49t): a new version of the program. Comput Phys Commun 183:166
24. Bardeen J, Cooper LN, Schrieffer JR (1957) Microscopic theory of superconductivity. Phys Rev 106:162
25. Bardeen J, Cooper LN, Schrieffer JR (1957) Theory of superconductivity. Phys Rev 108:1175
26. Chabanat E, Bonche P, Haensel P, Meyer J, Schaeffer R (1998) A Skyrme parametrization from subnuclear neutron star densities: part II. Nuclei far from stabilities. Nucl Phys A 635:231
27. Bartel J, Quentin Ph, Brack M, Guet C, Håkansson H-B (1982) Towards a better parametrisation of Skyrme-like effective forces: a critical study of the SkM force. Nucl Phys A 386:79
28. Lipkin HJ (1960) Collective motion in many-particle systems: part 1. The violation of conservation laws. Ann Phys 9:272–291
29. Nogami Y (1964) Improved superconductivity approximation for the pairing interaction in nuclei. Phys Rev 134:B313
30. Pradhan HC, Nogami Y, Law J (1973) Study of approximations in the nuclear pairing-force problem. Nucl Phys A 201:357–368
31. Otsuka T, Arima A, Iachello F, Talmi I (1978) Shell model description of interacting bosons. Phys Lett B 76:139
32. Otsuka T, Arima A, Iachello F (1978) Shell model description of interacting bosons. Nucl Phys A 309:1
33. Ginocchio JN, Kirson MW (1980) Relationship between the Bohr collective hamiltonian and the interacting-Boson model. Phys Rev Lett 44:1744
34. Dieperink AEL, Scholten O, Iachello F (1980) Classical limit of the interacting-Boson model. Phys Rev Lett 44:1747
35. Bohr A, Mottelson BR (1980) Features of nuclear deformations produced by the alignment of individual particles or pairs. Phys Scripta 22:468
36. Bohr A, Mottelson BR (1969, 1975) Nuclear structure, vol. I single-particle motion : vol. II nuclear deformations. Benjamin, New York.
37. Warner DD, Casten RF (1983) Predictions of the interacting boson approximation in a consistent Q framework. Phys Rev C 28:1798
38. Otsuka T (1993) Algebraic approaches to nuclear structure. In , Casten RF (ed) Harwood, Chur, p 195
39. Ginocchio JN, Kirson M (1980) An intrinsic state for the interacting boson model and its relationship to the Bohr-Mottelson model. Nucl Phys A 350:31
40. Otsuka T (1981) Microscopic basis of the proton-neutron interacting-Boson model. Phys Rev Lett 46:710
41. Otsuka T, Ginocchio JN (1985) Renormalization of g-Boson effects in the interacting-Boson hamiltonian. Phys Rev Lett 55:276
42. Rowe DJ, Iachello F (1983) Group theoretical models of giant resonance splittings in deformed nuclei. Phys Lett B 130:231
43. Elliott JP, White AP (1980) An isospin invariant form of the interacting boson model. Phys Lett B 97:169
44. Elliott JP, Evans JA (1981) An intrinsic spin for interacting bosons. Phys Lett B 101:216
45. Iachello F, Van Isacker P (1991) The interacting boson-fermion model. Cambridge University Press, Cambridge
46. Iachello F, Scholten O (1979) Interacting Boson-Fermion Model of collective states in odd-A nuclei. Phys Rev Lett 43:679
47. Brant S, Paar V, Vretenar D (1984) SU(6) Model for odd-odd nuclei (OTQM/IBOM) and Boson-Aligned phase diagram in 0(6) limit. Z Phys A 319:355
48. Iachello F (1980) Dynamical supersymmetries in nuclei. Phys Rev Lett 40:772
49. Jolie J, Heinze S, Van Isacker P, Casten RF (2004) Shape phase transitions in odd-mass nuclei using a supersymmetric approach. Phys Rev C 70:011305(R).

References

50. Otsuka T, Yoshida N (1985) User's manual of the program NPBOS, JAERI-M report 85 (Japan Atomic Energy Research Institute).
51. Belyaev ST, Zelevinski VG (1962) Anharmonic effects of quadrupole oscillations of spherical nuclei. Nucl Phys 39:582
52. Marumori T, Yamamura M, Tokunaga A, Takada K (1964) On the "Anharmonic Effects" on the collective oscillations in spherical even nuclei. I. Prog Theor Phys 31:1009
53. Scholten O, Iachello F, Arima A (1978) Interacting Boson Model of collective nuclear states III. The transition from SU(5) to SU(3). Ann Phys (N.Y.) 115:325.
54. Otsuka T (1984) "Independent-pair" property of condensed coherent pairs and derivation of the IBM quadrupole operator. Phys Lett B 138:1
55. Scholten O (1983) Microscopic calculations for the interacting boson model. Phys Rev C 28:1783
56. Iachello F (2001) Analytic description of critical point nuclei in a spherical-axially deformed shape phase transition. Phys Rev Lett 87:052501
57. Casten RF, Zamfir NV (2001) Empirical realization of a critical point description in atomic nuclei. Phys Rev Lett 87:052503
58. Nikšić T, Vretenar D, Lalazissis GA, Ring P (2007) Microscopic description of nuclear quantum phase transitions. Phys Rev Lett 99:092502
59. Nomura K, Shimizu N, Otsuka T (2010) Formulating the interacting boson model by mean-field methods. Phys Rev C 81:044307
60. Kaiser G (1994) A friendly guide to wavelets. Birkhauser, Boston
61. Brookhaven National Nuclear Data Center (NNDC) http://www.nndc.bnl.gov/nudat2/index.jsp
62. Rodriguez-Guzman RR, Egido JL, Robledo LM (2004) Beyond mean field description of shape coexistence in neutron-deficient Pb isotopes. Phys Rev C 69:054319
63. Börner HG, Mutti P, Jentschel M, Zamfir NV, Casten RF, McCutchan EA, Krüken R (2006) Low-energy phonon structure of ^{150}Sm. Phys Rev C 73:034314
64. Kulp WD, Wood JL, Allmond JM, Eimer J, Furse D, Krane KS, Loats J, Schmelzenbach P, Stapels CJ, Larimer R-M (2007) $N = 90$ region: the decays of ^{152}Eu^{m,g} to ^{152}Sm. Phys Rev C 76:034319
65. Bhat MR (2000) Nuclear data sheets for A = 148. Nucl. Data Sheets 89:797
66. Reich CW (2009) Nuclear data sheets for A = 154. Nucl. Data Sheets 110:2257
67. Torrence C, Compo GP (1998) A practical guide to wavelet analysis. Bull Am Meteorol Soc 79:61
68. Shevchenko A, Carter J, Cooper GRJ, Fearick RW, Kalmykov Y, von Neumann-Cosel P, Ponomarev V Yu, Richter A, Usman I, Wambach J (2008) Analysis of fine structure in the nuclear continuum. Phys Rev C 77:024302

Chapter 3
Rotating Deformed Systems with Axial Symmetry

3.1 A Piece of History, and Basics

In Chap. 2 it has been shown that, in the method of Ref. [1], the deformation energy surface of the self-consistent mean-field calculation is compared to the corresponding energy surface of IBM to obtain the parameters of IBM Hamiltonian. This procedure has turned out to be valid particularly for nuclei with weak to moderate quadrupole deformation, and has been practiced extensively in various mass region [2–5]. When a nucleus is well deformed, however, the rotational spectrum in actual nuclear system appears to be systematically different from the corresponding bosonic one (cf. Fig. 2.5). This is manifested by too small bosonic moment of inertia as compared to the corresponding fermionic, i.e., experimental, moment of inertia [1, 2].

This kind of difference has been known, in many cases, as a result of limited degrees of freedom in IBM, which in its standard version is comprised of s and d bosons only [6, 7]. In order to remedy this problem, other type of collective nucleon pairs, e.g., the $L = 4^+$ (G) pair, and the corresponding boson image (g boson) have been introduced, and their effects were renormalized into sd boson sector, yielding IBM Hamiltonians consistent with phenomenological ones [6–17]. Meanwhile, the validity of IBM for rotational nuclei was analyzed in terms of the Nilsson plus BCS model [18], coming up with the criticism that the SD-pair truncation may not be sufficient for describing intrinsic states of strongly deformed nuclei, and this naturally casts a question concerning the applicability of the IBM to rotational nuclei in particular. While it has been reported that the SD-pair dominance holds to a good extent in intrinsic states of rotational nuclei [13, 19, 20], there has been no conclusive mapping procedure from nucleonic systems to the corresponding IBM systems covering rotational nuclei. It is thus of much interest to revisit this issue with the newly proposed method of Ref. [1], looking for a prescription to cure the afore-mentioned problem of too small IBM moment of inertia.

In the method presented in Ref. [1], we calculated the energies of nucleonic and bosonic intrinsic states representing various quadrupole deformations, and obtained energy surfaces. We then determined parameters of the IBM Hamiltonian so that

Fig. 3.1 Concept of the rotational cranking. The intrinsic state for axially deformed system (indicated by the object with *broken line*) is rotated by the small angle. The ellipsoid whose boundary is indicated by the *solid line* represents the rotated intrinsic shape of the deformed system. The *arrows* going through both ellipsoids represent the symmetry axises of the intrinsic frame

the bosonic energy surface becomes identical to the nucleonic one [1], as shown in Sect. 2.4.1. These intrinsic states are at rest with rotational frequency $\omega = 0$. In this chapter, we move on by one step further with non-zero rotational frequency $\omega \neq 0$. Actually we analyze the responses of the nucleonic and bosonic intrinsic states by rotational cranking with infinitesimal ω. From such responses, one can extract the most important rotational correction to the IBM Hamiltonian.

Figure 3.1 illustrates the concept of the rotational cranking. The nucleon intrinsic state $|\Phi_F\rangle$ is obtained from the Hartree-Fock plus BCS (HF+BCS) calculation of the same type as the one done in Refs. [1, 2]. The Skyrme functional SkM* [21] is used throughout, while different types of the Skyrme functional do not alter the conclusion.

We start with the simple IBM-2 Hamiltonian, which is the same as the one used in the analyses in the Chap. 2, Eq. (2.41). Let us here recall the form of the Hamiltonian:

$$\hat{H}_B = \varepsilon \hat{n}_d + \kappa \hat{Q}_\pi \cdot \hat{Q}_\nu, \tag{3.1}$$

where the first and the second terms represent the d-boson number operator and the quadrupole-quadrupole interactions, respectively. These terms have already been defined in Eqs. (2.42) and (2.44), respectively. To make the difference between fermion and boson systems clearer, all operators, states and observables for fermion and boson systems in this chapter are characterized by the subscripts F (fermions) and B (bosons), respectively. Here the parameters ε, κ, χ_π and χ_ν are determined by comparing nucleonic and bosonic $\beta\gamma$ energy surfaces, following the method of Refs. [1, 2] described in Chap. 2, and are presented in Fig. 2.4g–l. The coherent state is denoted by $|\Phi_B\rangle$, which is exactly the same as the one that appeared already in Eq. (2.36).

3.2 Rotational Cranking

We now look into the problem of rotational response. We shall restrict ourselves to nuclei with axially symmetric strong deformation, because this problem is crucial to those nuclei but is not so relevant to the others. An axially symmetric intrinsic state is invariant with respect to the rotation around the symmetry (z) axis. This

3.2 Rotational Cranking

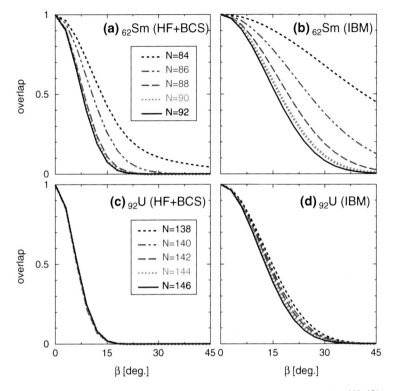

Fig. 3.2 Overlap between the intrinsic state and the rotated one at angle β for $^{146-154}$Sm and $^{230-238}$U nuclei for **a, c** fermion (HF+BCS) and **b, d** boson (IBM) systems. Skyrme SkM* functional is used. The figure has been taken from Ref. [22]

means $a_{\rho\mu} = 0$ for $\mu \neq 0$ in Eq. (2.36) in the case of bosons. Such intrinsic states of nucleons and bosons are supposed to be obtained as the minima of the energy surfaces. Let us now rotate the axially symmetric intrinsic states about the y-axis by angle β. Figure 3.2 shows the overlap between the intrinsic state $|\Phi_X\rangle$ and the rotated one $|\Phi'_X\rangle = e^{-i\hat{L}_y\beta}|\Phi_X\rangle$, where X stands for either fermion ($X = F$) or boson ($X = B$). Here $\hat{L}_y$ denote the y-component of the angular momentum operator for boson or nucleon system. We take $^{146-154}$Sm and $^{230-238}$U nuclei as examples. Some of these nuclei are good examples of SU(3) limit of IBM [23].

Figure 3.2a, c, as well as Fig. 3.2b, d, shows the overlaps for nucleons and bosons,[1] respectively. For Sm isotopes, the parameters of the boson Hamiltonian $\hat{H}_B$ are taken from [2], which have been presented also in Fig. 2.4g–l, while the parameters for U isotopes are determined in the same way as $\varepsilon \approx 0.100$ MeV, $\kappa \approx -0.18$ MeV, and $\chi_\pi \approx \chi_\nu \approx -1.0$, which characterize the deformed nuclei close to the SU(3) limit [24]. These parameters for Sm and U isotopes are used throughout this chapter.

---

[1] Calculation of the overlap is described in Appendix B.1.

In each case of Fig. 3.2, the overlap is peaked at the rotation angle of $\beta = 0°$ with the value unity, and decreases with β. The nucleonic overlaps are peaked more sharply, whereas boson ones are damped more slowly. It is clear that as a function of β, boson rotated intrinsic state changes more slowly than the corresponding nucleon one, due to limited degrees of freedom for IBM consisting of s and d bosons only.

We point out that the overlap becomes narrower in β with the neutron number N for Sm isotopes (see Fig. 3.2a, b). This is related to the growth of deformation. On the other hand, there is no notable change in the overlap for these U isotopes, because pronounced prolate minimum appears always at $\beta_2 \sim 0.25$ in their energy surface, with β_2 denoting the axially-symmetric deformation in the geometrical model.

The nucleon-boson difference of the rotational response discussed so far suggests that the rotational spectrum of a nucleonic system may not be fully reproduced by the boson system determined by the mapping method of Ref. [1] using the energy surfaces at rest. In fact, it will be shown later that the moment of inertia of a nucleon system differs from the one calculated by the mapped boson Hamiltonian. We then propose to introduce a term into the boson Hamiltonian, so as to keep the energy surface-based mapping procedure but incorporate the different rotational responses. This term takes the form of $\hat{L} \cdot \hat{L}$ where $\hat{L}$ denotes the boson angular momentum operator:

$$\hat{L} = \hat{L}_\pi + \hat{L}_\nu \quad \text{with} \quad \hat{L}_\rho = \sqrt{10}[d_\rho^\dagger \tilde{d}_\rho]^{(1)}. \tag{3.2}$$

This term is nothing but the squared magnitude of the angular momentum with the eigenvalue $L(L+1)$, and changes the moment of inertia of rotational band keeping their wave functions. A phenomenological term of this form was used in the fitting calculation of IBM, particularly in its SU(3) limit [23], without knowing its origin nor physical significance.

We adopt, hereafter, a Hamiltonian, $\hat{H}'_B$, which includes this term with coupling constant α:

$$\hat{H}'_B = \hat{H}_B + \alpha \hat{L} \cdot \hat{L}, \tag{3.3}$$

where $\hat{H}_B$ is given in Eq. (3.1). The $\alpha \hat{L} \cdot \hat{L}$ term will be referred to as LL term hereafter. The LL term contributes to the energy surface in the same way as a change of d-boson energy $\Delta\varepsilon = 6\alpha$ (see Eq. (3.1)), because the energy surface at rest (i.e., $\omega = 0$) is formed by the boson intrinsic state $|\Phi_B\rangle$ containing no $d_{\pm 1}$ component. Hence, by shifting ε slightly, we obtain the same energy surface as the one without the LL term, and consequently the other parameters of mapped H_B remain unchanged.

We now turn to the determination of α in Eq. (3.3). First, we perform the cranking model calculation for the fermion system to obtain its moment of inertia, denoted by $\mathscr{J}_F$, in the usual way [26]. By taking the Inglis-Belyaev (IB) formula, we obtain [27, 28]

$$\mathscr{J}_F = 2 \cdot \sum_{i,j>0} \frac{|\langle i|\hat{L}_k|j\rangle|^2}{E_i + E_j}(u_i v_j - u_j v_i)^2, \tag{3.4}$$

3.2 Rotational Cranking

where energy E_i and v-factor v_i of quasi-particle state i are calculated by the HF+BCS method of Refs. [29, 30]. Here, L_k is the nucleon angular momentum operator, and k means the axis of the cranking rotation, being either x or y, as z-axis. Based on the earlier argument, the y-axis is chosen following a conventional notation.

Next, the bosonic moment of inertia, denoted as $\mathscr{J}_B$, is calculated by the cranking formula of Ref. [31] with $d_{\pm 1}$ being mixed, to an infinitesimal order, into the coherent state $|\Phi_B\rangle$ of Eq. (2.36):

$$\mathscr{J}_B = \lim_{\omega \to 0} \frac{1}{\omega} \frac{\langle \Phi_B | \hat{L}_k | \Phi_B \rangle}{\langle \Phi_B | \Phi_B \rangle}, \tag{3.5}$$

with

$$|\Phi_B\rangle \propto \prod_{\rho=\pi,\nu} \left\{ s_\rho + \beta_B \cos \gamma_B d^\dagger_{\rho 0} + \frac{1}{\sqrt{2}} \sin \gamma_B (d^\dagger_{\rho+2} + d^\dagger_{\rho-2}) \right.$$
$$\left. + a_{\rho 1} d^\dagger_{\rho+1} + a_{\rho-1} d^\dagger_{\rho-1} \right\}^{N_\rho} |0\rangle \tag{3.6}$$

where ω is the cranking frequency, $a_{\rho+1}$ ($=a_{\rho-1}$) denotes the amplitude for $d_{\rho\pm 1}$ bosons, and $\hat{L}_k$ stands for the k-component of boson angular momentum operator. Note that $a_{\rho\pm 1} \propto \omega$ at the limit of $\omega \to 0$, leading $\mathscr{J}_B$ to a finite value. The exact form of the IBM cranking moment of inertia can be found in Appendix B.1.

The value of α is determined for individual nucleus so that the corresponding bosonic moment of inertia, $\mathscr{J}_B$ in Eq. (4.9) becomes equal to $\mathscr{J}_F$ in Eq. (3.4). This prescription makes sense, if the nucleus is strongly deformed and the fixed intrinsic state is so stable as to produce individual levels of a rotational band through the angular momentum projection in a good approximation. The resultant excitation energies should follow the rotor formula $E_x \propto L(L+1)$ for L being the angular momentum of the level. The present prescription with the LL term should be applied only to certain nuclei which belong to this type. We introduce a criterion to select such nuclei in terms of the ratio $R_{4/2} = E_x(4_1^+)/E_x(2_1^+)$, and set a minimum value for this. Empirical systematics [33] suggests that the evolution towards stronger deformation continues as the number of valence nucleons increases, but this evolution becomes saturated beyond $R_{4/2} \sim 3.2$. Namely, for the nuclei with $R_{4/2} > 3.2$, the deformation is considered to be evolved sufficiently well, and we take $R_{4/2} > 3.2$ as the criterion to apply the LL term. This discrete criterion is also for the sake of simplicity, but the major discussions of this work do not depend on its details.

Figure 3.3a–c shows the moments of inertia for Sm and U isotopes. In these figures, $\mathscr{J}_B$ calculated with the LL term (w/ LL), $\mathscr{J}_B$ calculated without it (w/o LL), and $\mathscr{J}_F$ are compared. Experimental ones determined from the 2_1^+ levels [32] are shown also.

We divide Sm isotopes into two categories according to the criterion defined above. First, the ratio $R_{4/2}$ is calculated without the LL term, leading to $^{152-158}$Sm with $R_{4/2} > 3.2$ and $^{146-150}$Sm with $R_{4/2} < 3.2$. For the former category, the LL

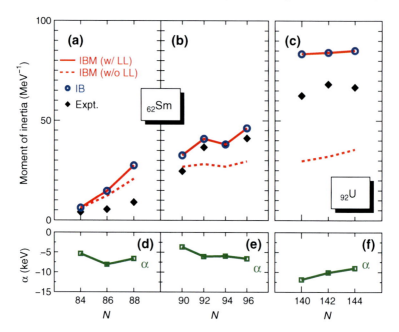

Fig. 3.3 (*Upper panels*) Moments of inertia in the intrinsic state for **a** $^{146-150}$Sm, **b** $^{152-158}$Sm and **c** $^{232-236}$U, calculated by IBM with (w/) and without (w/o) the LL term and by Inglis-Belyaev formula. Experimental data taken from 2_1^+ excitation energies [32] are also shown. (*Lower panel*) The derived α value for **d** $^{146-150}$Sm, **e** $^{152-158}$Sm and **f** $^{232-236}$U as a function of neutron number N. Skyrme functional SkM* is used. The figure has been taken from Ref. [22]

term should be included, and Fig. 3.3b demonstrates that the LL term has significant effects so as to be consistent with the experimental moment of inertia. To be more precise, the experimental value is relatively large for $N \geq 90$, and looks nearly flat for the nuclei with $N \geq 92$, being $35 \sim 40$ MeV^{-1}. Enlargement of the moment of inertia means that the parameter α should be negative. The IB formula reproduces this trend quite well, which is inherited to bosons by the present method.

Although the LL term should not be used for the category depicted in Fig. 3.3a, one can observe some features that both $\mathscr{J}_F$ and $\mathscr{J}_B$ increase with N. Although experimental moment of inertia exhibits a gap between Fig. 3.3a, b (from $N = 88$ to 90), neither $\mathscr{J}_B$ nor $\mathscr{J}_F$ follow this trend, showing only gradual changes. This could be, for example, due to the absence of the particle number conservation in the Skyrme EDF calculation. We do not touch on this point in this chapter.

Figure 3.3d, e shows, respectively, the derived α value for the $^{146-150}$Sm and the $^{152-158}$Sm nuclei. First we notice an overall trend that α does not change so much, while the IB value of $\mathscr{J}_F$ changes by an order of magnitude. Although the α values for the $^{146-150}$Sm nuclei do not make much sense, this is of certain interest.

Figure 3.3c shows the moments of inertia for the $^{232-236}$U nuclei, which are rather flat. We point out that the calculated moment of inertia, $\mathscr{J}_F = \mathscr{J}_B$ with the LL

3.2 Rotational Cranking

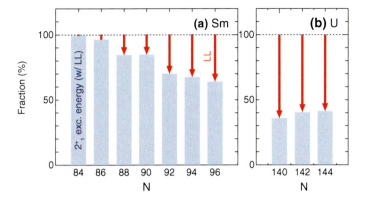

Fig. 3.4 Fraction of the 2_1^+ excitation energy $E_x(2_1^+)$ that includes the LL term, normalized with respect to the excitation energy $E_x(2_1^+)$ that does not include the LL term, for **a** Sm and **b** U isotopes. *Downward arrows* represent the LL matrix elements normalized by $E_x(2_1^+)$ (denoted by "w/o LL"). The figure has been taken from Ref. [22]

term, turned out to be about twice large as that of $^{152-158}$Sm. This dramatic change is consistent with experiment, although somewhat overshoots experimental changes.

We shall then discuss eigenvalues of $\hat{H}'_B$ in Eq. (3.3) obtained by the diagonalization using NPBOS code [34]. We first investigate to what extent $E_x(2_1^+)$ is lowered by the LL term. Figure 3.4 shows the fraction of this lowering, by normalizing it with respect to the $E_x(2_1^+)$ without the LL term, for (a) Sm and (b) U isotopes. This lowering is, as indicated by the arrows in Fig. 3.4, >30 % for $^{154-160}$Sm and >60 % for $^{232-236}$U. On the other side, it is almost vanished or quite small for $N = 84$–90. Thus, it may not affect the IBM description much, even if one keeps the LL term in all nuclei. We do not take it, because the present derivation does not give physical basis for the LL term for nuclei without strong deformation.

3.3 Results and Discussions

3.3.1 Rotational Bands

Figure 3.5 shows the evolution of low-lying yrast spectra for (a) Sm and the neighboring (b) Gd isotopes as functions of N. For both Sm and Gd isotopes, the LL term is included for $N \geq 90$, but is not included for $N \leq 88$, based on the criterion discussed above. The IBM parameters for Gd isotopes are derived similarly to those used for Sm isotopes. Figure 3.5a, b indicates that calculated spectra become more compressed with N and exhibit rotational feature for $N \geq 90$, similarly to the experimental trends [32]. One notices a certain deviation at $N = 88$, where the Skyrme energy surface favors stronger deformation and the calculated excitation energies are somewhat too low [2].

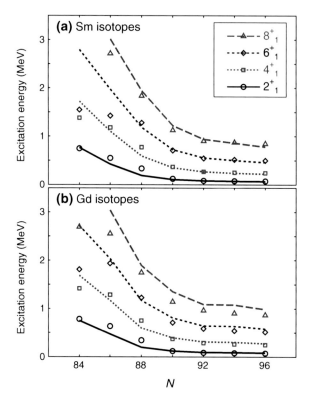

Fig. 3.5 Experimental [32] (*symbols*) and calculated (*curves*) yrast spectra for **a** Sm and **b** Gd isotopes as functions of neutron number N. Skyrme functional SkM* is used. The figure has been taken from Ref. [22]

Figure 3.6 shows yrast levels of ^{154}Sm, ^{156}Gd, ^{230}Th and ^{230}U nuclei as representatives of rotational nuclei. The LL term is included for these nuclei, as they fulfill the criterion. For the ^{230}Th nucleus, the parameters of the Hamiltonian $\hat{H}_B$ take almost the same values as those for the ^{232}U nucleus. A nice overall agreement arises between the theoretical and the experimental [32] spectra, and the contribution of the LL term to it is remarkable. Particularly for the ^{154}Sm and the ^{230}Th nuclei, the calculated spectra look nearly identical to the experimental ones.

We now comment on side-band levels. The deviations of β-bandhead (0_2^+) and γ-bandhead (2_2^+) energies are improved by tens of keV by the LL term. However, these band-head energies are still much higher than experimental ones. Thus, there are still open questions on side-band levels. Nevertheless, the relative spacing inside the bands is reduced by hundreds of keV, producing certain improvements.

We here remind the reader of those studies which attempt to derive a general collective Hamiltonian from a given EDF, where the self-consistent mean-field energy surface supplemented with zero point rotational and vibrational corrections are treated as a collective potential [35–39]. Generalized kinetic energy terms for both rotational

3.3 Results and Discussions

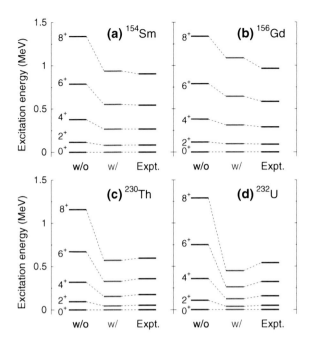

Fig. 3.6 Level schemes of typical strongly deformed rotational nuclei close to the SU(3) limit of IBM: **a** ^{154}Sm, **b** ^{156}Gd, **c** ^{230}Th and **d** ^{232}U nuclei. Calculated spectra with (denoted by w/) and without (denoted by w/o) the LL term, are compared with experimental [32] spectra. Skyrme functional SkM* is used. The figure has been taken from Ref. [22]

and vibrational motions come out in such approaches. In the present work, we compare the results of Skyrme EDF with the corresponding results of the mapped boson system, at the levels of the energy surface and the rotational response. The kinetic energies of nucleons are included in both levels, while the rotational kinetic-like boson term appears from the latter.

3.3.2 Validity of Cranking Formula

Before closing this chapter, what should be worth mentioning is the validity of the cranking (Inglis-Belyaev) moment of inertia. This question naturally arises because it is generally known that, when incorporated into a collective Hamiltonian approach (e.g., [39]), the Inglis-Belyaev formula underestimates the mass parameters of the kinetic energies, leading to the systematic deviations of the rotational band. In the framework of the generator coordinate method (GCM), collective mass can be deduced with the Gaussian overlap approximation (GOA). The mass parameter in the GOA is often replaced with the cranking mass. The shortcoming of the Inglis-Belyaev formula has been ascribed, in the framework of the pairing plus quadrupole model [40, 41], to the fact that the time-odd component (that breaks the time reversal invariance) arising from the moving mean field is ignored in the cranking approximation. Effect of the time-odd component may become large in reproducing the yrast states of deformed nuclei when it is incorporated into the cranking mass parameters obtained from a modern energy density functional, and the collective Hamiltonian is derived using this corrected cranking masses [42]. In the present work, as shown in Fig. 3.6, almost perfect agreement is achieved even though the Inglis-Belyaev formula is used in determining the parameter of the LL term. The role of the cranking moment of inertia is rather different between the present and the collective Hamiltonian approaches: In the former, only the difference between the fermion and the boson intrinsic (cranking) moments of inertia is of relevance in order to fix the LL coefficient. As the boson intrinsic moment of inertia is in many cases sufficiently smaller than the fermion one in deformed nuclei, the particular deficiency of the cranking approximation itself may not severely matter the resultant rotational band. In the latter approach, however, the cranking moment of inertia is used explicitly as mass parameter, which could alter the final result to a large extent. In the IBM system, necessary dynamical effects could be in principle included through the diagonalization of the Hamiltonian. Nevertheless, anything definite cannot be concluded at the moment concerning the extent to which the time-odd field is included in the boson system quantitatively, because one cannot obviously make any one-to-one correspondence from the time-odd field in the nucleon system to an appropriate one in the boson system.

3.4 Brief Summary

To summarize this chapter, we have proposed a novel formulation of the IBM for rotational nuclei. The rotation of strongly deformed multi-nucleon system differs, in its response to the rotational cranking, from its boson image obtained by the mapping method of Ref. [1] where the energy surface at rest is considered. Significant differences then appear in moment of inertia between nucleon and boson systems. We have shown that this problem is remedied by introducing the LL term into the IBM Hamiltonian. The effect of the LL term makes essential contribution to rotational spectra, solving the longstanding problem of too small moment of inertia microscopically. Experimental data are reproduced quite well, without any phenomenological adjustment. The mapping of Ref. [1] appears quite sufficient for vibrational and γ-unstable nuclei, and the present study makes the IBM description of strongly deformed nuclei sensible theoretically and empirically. Thus, we seem to have come to the stage of having microscopic basis of the IBM in all situation at the lowest order. Meanwhile, this achievement is partly due to the successful description of Skyrme energy density functional. The feature discussed in this chapter is related to the question as to whether the IBM can be applied to deformed nuclei or not [18]. The present work indicates that the rotational response is substantially different between fermions and bosons, but the difference can be incorporated into the IBM in a microscopic and self-consistent way.

References

1. Nomura K, Shimizu N, Otsuka T (2008) Mean-field derivation of the Interacting Boson model Hamiltonian and exotic nuclei. Phys Rev Lett 101:142501
2. Nomura K, Shimizu N, Otsuka T (2010) Formulating the interacting Boson model by mean-field methods. Phys Rev C 81:044307
3. Nomura K, Otsuka T, Rodríguez-Guzmán R, Robledo LM, Sarriguren P (2011) Structural evolution in Pt isotopes with the interacting Boson model Hamiltonian derived from the Gogny energy density functional. Phys Rev C 83:014309
4. Nomura K, Otsuka T, Rodríguez-Guzmán R, Robledo LM, Sarriguren P, Regan PH, Stevenson PD, Zs. Podolyák (2011) Spectroscopic calculations of the low-lying structure in exotic Os and W isotopes. Phys Rev C 83:051303
5. Nomura K, Otsuka T, Rodríguez-Guzmán R, Robledo LM, Sarriguren P (2011) Collective structural evolution in Yb, Hf, W, Os and Pt isotopes. Phys Rev C 84:054316
6. Otsuka T (1979) Boson model of medium-heavy nuclei. Ph.D thesis, University of Tokyo
7. Otsuka T (1981) Rotational states and interacting bosons. Phys Rev Lett 46:710
8. Mizusaki T, Otsuka T (1996) Microscopic calculations for O(6) nuclei by the interacting Boson model. Prog Theor Phys Suppl 125:97–150
9. Scholten O (1983) Microscopic calculations for the interacting Boson model. Phys Rev C 28:1783
10. Otsuka T, Yoshinaga N (1986) Fermion-Boson mapping for deformed nuclei. Phys Lett B 168:1
11. Otsuka T (1984) "Independent-pair" property of condensed coherent pairs and derivation of the IBM quadrupole operator. Phys Lett B 138:1
12. Otsuka T, Ginocchio JN (1985) Renormalization of g-Boson effects in the interacting-Boson Hamiltonian. Phys Rev Lett 55:276
13. Otsuka T (1981) Rotational states and interacting bosons. Nucl Phys A368:244
14. Yoshinaga N, Arima A, Otsuka T (1984) A microscopic approach to a foundation of the interacting Boson model by using the interacting Boson model by using angular momentum projection. Phys Lett B 143:5
15. Zirnbauer MR (1984) Microscopic approach to the interacting Boson model (II). Extension to neutron-proton systems and applications. Nucl Phys A 419:241
16. Pannert W, Ring P, Gambhir YK (1985) An analysis of angular-momentum projected Hartree-Fock-Bogoliubov wave functions in terms of interacting bosons. Nucl Phys A 443:189
17. Otsuka T, Sugita M (1988) Proton-Neutron sdg Boson model and spherical-Deformed phase Transition. Phys Lett B 215:205
18. Bohr A, Mottelson BR (1980) Features of nuclear deformations produced by the alignment of individual particles or pairs. Phys Scripta 22:468
19. Otsuka T, Arima A, Yoshinaga N (1982) Dominance of monopole and quadrupole pairs in the Nilson model. Phys Rev Lett 48:387
20. Bes DR, Broglia RA, Maglione E, Vitturi A (1982) Nilson and interacting-Boson model pictures of deformed nuclei. Phys Rev Lett 48:1001
21. Bartel J, Quentin Ph, Brack M, Guet C, Håkansson H-B (1982) Towards a better parametrisation of Skyrme-like effective forces: A critical study of the SkM force. Nucl Phys A 386:79
22. K. Nomura, T. Otsuka, N. Shimizu, L. Guo (2011) Microscopic formulation of the interacting Boson model for rotational nuclei. Phys Rev C 83:041302(R)
23. Arima A, Iachello F (1978) Interacting Boson model of collective states:II. The rotational limit. Ann Phys 111:201–238
24. Arima A, Iachello F (1975) Collective nuclear states as representations of a SU(6) group. Phys. Rev. Lett. 35:1069
25. Iachello F, Arima A (1987) The interacting Boson model. Cambridge University Press, Cambridge
26. Ring P, Schuck P (1980) The nuclear many-body problem. Springer, Berlin
27. Inglis DR (1956) Nuclear moments of inertia due to nucleon motion in a rotating well. Phys Rev 103:1786

28. Belyaev ST (1961) Concerning the calculation of the nuclear moment of inertia. Nucl Phys 24:322
29. Guo L, Maruhn JA, Reinhard P-G (2007) Triaxiality and shape coexistence in germanium isotopes. Phys Rev C 76:034317
30. Guo L, Maruhn JA, Reinhard P-G, Hashimoto Y (2008) Conservation properties in the time-dependent Hartree Fock theory. Phys Rev C 77:041301(R)
31. Schaaser H, Brink DM (1984) Calculations away from SU(3) symmery by cranking the interacting Boson model. Phys Lett B 143:269
32. Brookhaven national nuclear data center (NNDC) http://www.nndc.bnl.gov/nudat2/index.jsp
33. Cakirli RB, Casten RF (2006) Direct empirical correlation between proton-neutron interaction strengths and the growth of collectivity in nuclei. Phys Rev Lett 96:132501
34. Otsuka T, Yoshida N (1985) User's manual of the program NPBOS, JAERI-M report 85, Japan Atomic Energy Research Institute
35. Bonche P, Dobaczewski J, Flocard H, Heenen P-H, Meyer J (1990) Analysis of the generator coordinate method in a study of shape isomerism in ^{194}Hg. Nucl Phys A 510:466
36. Delaroche J-P, Girod M, Libert L, Goutte H, Hilaire S, Peru S, Pillet N, Bertsch GF (2010) Structure of even-even nuclei using a mapped collective Hamiltonian and the D1S Gogny interaction. Phys Rev C 81:014303
37. Nikšić T, Li ZP, Vretenar D, Próchniak L, Meng J, Lalazissis GA, Ring P (2009) Beyond the relativistic mean-field approximation III. Collective Hamiltonian in five dimensions. Phys Rev C 79:034303
38. Li ZP, Nikšić T, Vretenar D, Meng J, Lalazissis GA, Ring P (2009) Microscopic analysis of nuclear quantum phase transitions in the N≈90 region. Phys Rev C 79:054301
39. Li ZP, Nikšić T, Vretenar D, Meng J (2010) Microscopic description of spherical to γ-soft shape transitions in Ba and Xe nuclei. Phys Rev C 81:034316
40. Balyaev ST (1965) Time-dependent self-consistent field and collective nuclear Hamiltonian. Nucl Phys 64:17
41. Baranger M, Vénéroni M (1978) An adiabatic time-dependent Hartree-Fock theory of collective motion in finite systems. Ann Phys 114:123
42. Vretenar D (2011) private communication

Chapter 4
Weakly Deformed Systems with Triaxial Dynamics

4.1 Quantum Phase Transitions

Before discussing each particular case, we review in this section the concepts of quantum/shape phase transition and of the critical-point symmetries as they are useful to better interpret the shape phenomena considered in this chapter.

Quantum phase transition (QPT) is one of the central issues in finite quantal systems, including atomic nucleus and other mesoscopic systems, as well as in high-energy and condensed matter physics. Particularly, equilibrium nuclear shape, e.g., of quadrupole-type, undergoes the distinct structural evolution between spherical vibrational and deformed rotational states with Z and/or N. Here the nuclear QPT in this context means the one that occurs at a specific number of N and/or Z, which is discrete and as such should differ from the usual phase transition of thermodynamic type. As the QPT in nuclei is rather unique, it has drawn much attention from various perspectives (for review, see Refs. [1, 2], for instance).

Figure 4.1 depicts the *phase diagram* (or, the IBM-1 symmetry triangle), where each of the vertices corresponds to a limit of three dynamical symmetries of IBM. The diagram is drawn in terms of only two strength parameters of the IBM-1 Hamiltonian in the consistent-Q formalism, which is similar to the one of Ising type:

$$\hat{H}_{\text{CQF}} = \varepsilon \hat{n}_d - \kappa \hat{Q} \cdot \hat{Q}$$
$$= c\left[(1-\zeta)\hat{n}_d - \frac{\zeta}{4N_B}\hat{Q}\cdot\hat{Q}\right] \text{ with } \zeta = \frac{4N_B}{4N_B + \varepsilon/\kappa}, \qquad (4.1)$$

where c and N_B denote the overall scaling factor irrelevant to the QPT and the number of bosons, respectively. The d-boson number operator $\hat{n}_d = d^\dagger \cdot \tilde{d}$ and the quadrupole operator $\hat{Q} = d^\dagger s + s^\dagger \tilde{d} + \chi [d^\dagger \tilde{d}]^{(2)}$ correspond to their IBM-2 analogues in Eqs. (2.42) and (2.44), respectively. The equilibrium shape of a given nucleus can be specified by the two parameters ζ and χ: ζ is related to the ratio ε/κ and controls the competition between the spherical-driving ($\hat{n}_d$) term and the deformation-driving ($\hat{Q}\cdot\hat{Q}$) term. χ has the similar meaning to the ones in IBM-2 (cf. Eq. (2.44)) and

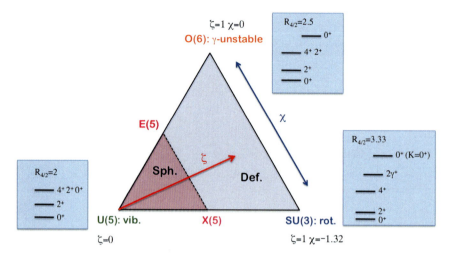

Fig. 4.1 Symmetry triangle of IBM-1, or equivalently the phase diagram with respect to the parameters in the Hamiltonian of Eq. (4.1): ζ, which corresponds to the ratio ε/κ and controls the transition between spherical and deformed phases, and χ, which determines the softness in γ. The *three vertices* represent the dynamical symmetries of the IBM. For each limit, level scheme is depicted together with the $R_{4/2}$ value, which denotes the ratio of the 4^+_1 to the 2^+_1 excitation energies. Note that oblate deformation and the transition between prolate and oblate deformed states as well are not considered here. According to the ζ and the χ values for a given nucleus, which are usually taken from the phenomenological fit, can one associate the nucleus with a specific point inside the symmetry triangle. The E(5) and X(5) are critical points for the U(5)-O(6) and U(5)-SU(3) transitions, respectively. The line connecting E(5) with X(5) separates the deformed and the spherical phases

determines the type of the deformation, i.e., the γ softness. According to the ζ and the χ values for a given nucleus, which are usually taken from the phenomenological fit, can one associate the nucleus with arbitrary point inside the symmetry triangle.[1] Along the lines of U(5)-SU(3) and U(5)-O(6) transitions in the symmetry triangle, one sees the X(5) [3] and E(5) [4] critical-point symmetries, respectively, which will be discussed below.

Order parameters of the QPT usually correspond to level energies of some low-lying states like the 2^+_1 state, electromagnetic transition rates, binding energies, two-nucleon separation energies ..., etc., which change discontinuously at a specific value of control parameter. The transition from U(5) to SU(3) symmetries and the one from U(5) to O(6) represent the first- and the second-order phase transitions, respectively, because it is empirically known that the order parameters and their derivative with respect to a control parameter can change discontinuously at particular number of nucleons.

---

[1] For instance, $\zeta \to 0$ ($\kappa \approx 0$) in U(5) limit; $\zeta \to 1$ ($\varepsilon \approx 0$) and $\chi \to -\sqrt{7}/2 (= -1.32)$ in SU(3); $\zeta \to 1$ ($\varepsilon \approx 0$) and $\chi \to 0$ in O(6). The ζ and χ values in each limit is depicted in Fig. 4.1.

4.1 Quantum Phase Transitions

We here make a distinction between the critical-point and the dynamical symmetries. Simply speaking, the former is for fermion and the latter boson systems: the X(5) and the E(5) symmetries represent the analytical solution of the geometrical collective Hamiltonian and have little to do with algebraic aspect [3, 4], while the dynamical symmetries U(5), SU(3) and O(6) are realized in the boson algebra. In the IBM, one can describe the critical-point and the dynamical symmetries in a unified manner by taking the classical limit of the IBM Hamiltonian in the coherent state. For the transition from the spherical to the axially deformed shapes in the general geometrical collective model, the collective potential energy surface should undergo the change in its topology from harmonic oscillator to the curve having a minimum at $\beta > 0$ as a function of valence nucleon number (see Fig. 2.1). The X(5) critical point possesses the potential energy surface taking on the feature that is totally flat in β direction and that behaves quadratically in γ. For the γ-unstable system, the collective potential is completely flat in γ. Then, in the transition from the spherical vibrational to γ-unstable deformed systems, the collective potential of the E(5) symmetry corresponds to the infinite square well in β direction but does not depend on γ. The classical limits of the IBM for the three-dynamical symmetries are provided in the analytical expressions, which are quite alike the harmonic oscillator for U(5), the one with sharp minimum at the finite value of β for SU(3), and the one that is totally flat in γ direction for O(6). Potentials similar to those in E(5) and X(5) models can also be realized in the boson model. Strictly speaking, however, the classical limit of the IBM Hamiltonian represents total energy, which should contain the effects of kinetic energies in addition to potential term.

The level schemes of both the X(5) and E(5) models can be obtained by solving the collective Hamiltonian under the infinite square-well potential with and without the γ dependence, respectively. With the assumptions of the potentials, one is led to the five dimensional differential equation, whose solutions can be obtained as Bessel functions. Quantum numbers of the X(5) and the E(5) symmetries correspond to zeros of the Bessel functions. These level schemes obtained from X(5) and E(5) models are completely parameter free but for the overall scale factors that are irrelevant to the qualitative studies.

Possible empirical evidence for the critical points has been studied for each transitional class and the critical-point symmetries. For example, in axially deformed nuclei in rare-earth region, drastic shape change as a function of neutron number is observed at the $N = 90$ isotones. This is recognized as the first order QPT from U(5) to SU(3) limits, and $N = 90$ nuclei can be the good examples of X(5) symmetry [5]. On the other hand, $A \sim 130$ region nuclei such as Xe-Ba nuclei are typical nuclei with significant γ instability and the U(5)-O(6) transition, where in particular ^{132}Ba and ^{134}Ba are nice examples of the O(6) and the E(5) symmetries [6]. It is much more difficult to find out the QPT along O(6)-SU(3), since one cannot identify any clear signature in an order parameter.

Other classes of phase transitions can be formulated when the symmetry triangle of Fig. 4.1 is extended. If one extrapolates the leg that starts from SU(3) to O(6) to go further beyond the O(6) vertex, another symmetry limit associated with the oblate axially-symmetric deformation (so-called $\overline{\text{SU(3)}}$ symmetry) can be defined. In this case, the O(6) can be viewed as the critical point locating in between prolate and oblate equilibrium shapes [7].

The concepts of QPT and critical-point symmetries are so oversimplified that it is not obvious whether such schematic descriptions apply to realistic nuclear systems in which considerable amount of quantum fluctuation enters. The following six sections are mainly devoted to this issue in comparison to the available experimental data and to other nuclear structure models.

In all the nuclei considered in this chapter, except for some Yb, Hf, W and Os nuclei near midshell (Sect. 4.3.3), no such problem arises that concerns the overall scale of the moment of inertia of rotational band, which was analyzed in Chap. 3. The use of the essential IBM Hamiltonian of Eq. (2.41) then suffices, which is derived only through the energy-surface analyses.

The thorough investigation of the first-order QPT in rare-earth region remains to be done along the same line, where rotational correction discussed in Chap. 3 is necessary. This issue has been addressed in [8].

4.2 Axially to γ-Unstable Deformed Nuclei

The nuclei with mass $A \approx 100$–130 exhibit very rich shape phenomena, including the γ-softness. A lot of phenomenological IBM calculations have been carried out in the past and have turned out to be valid for the nuclei in these mass regions because there are relatively enough experimental data to compare. More recently several possible evidence for the transition from the vibrational to the γ-soft shapes together with E(5) critical points have been pointed out in Ru and Ba isotopes.

The IBM Hamiltonian of the considered nuclei are derived using a unified way based on the EDF-to-IBM mapping procedure with a single parametrization of Skyrme functional SkM* in the next three subsections (Sects. 4.2.1–4.2.3), where most of the considered nuclei have modest deformation. Note that the qualitative features of the results and their overall agreement with the experimental data do not depend too much on the choice of the EDFs and on the pairing properties. Self-consistent constrained mean-field calculations presented in Sects. 4.2.1–4.2.3 have been performed by the HF+BCS method using ev8 code [9] with the Skyrme functional SkM* [10] throughout, similarly to the case in Sect. 2.4.1.

As the first example we consider the Ru and Pd isotopes with mass $A \approx 100$–120, where the typical spherical-to-γ-unstable shape transition has been observed.

4.2.1 $Z < 50$, $50 \leq N \leq 82$ Major Shells

Energy Surface

We show in Fig. 4.2 the energy surfaces for Ru isotopes for $N = 54$–80. The pattern of the energy surface changes with N moderately. For $N = 54$–62, the self-consistent HF+BCS (indicated by HF in Fig. 4.2) energy surface suggests a nearly spherical structure, which is slightly prolate deformed. The flat area in the β–γ plane becomes larger from $N = 62$ to 64 significantly, which suggests the transition from nearly spherical to γ-unstable shapes. The HF+BCS energy surface of Ru isotopes exhibits a weak triaxial deformation for $N = 64$–70, which is described by the flat IBM energy surface with $\chi_\pi + \chi_\nu \sim 0$. As seen in Fig. 4.2, the HF+BCS energy surface for $N = 64$–70 is quite complicated in topology so that it behaves like an

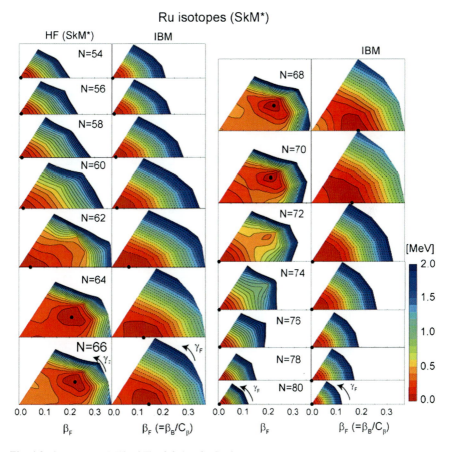

Fig. 4.2 Same as the RHS of Fig. 2.3, but for Ru isotopes

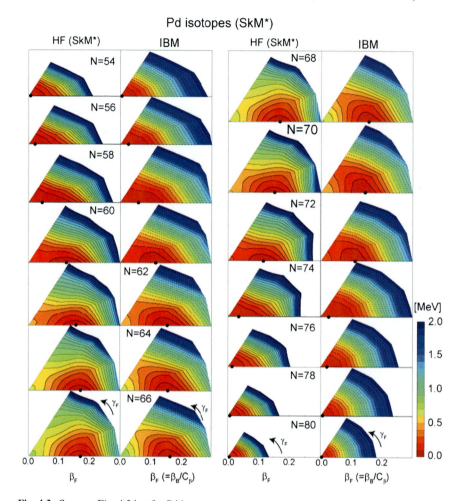

Fig. 4.3 Same as Fig. 4.2 but for Pd isotopes

infinite wall. This cannot be reproduced by the IBM energy surface of Eq. (2.46) and is far beyond the limit of the present energy surface fit.

The energy surfaces for Pd isotopes are depicted in Fig. 4.3. The self-consistent HF+BCS energy surface exhibits spherical structures in the vicinity of the shell closures $N = 50$ and 82, and shows a shallow prolate minimum for the open-shell region, without any notable change of the degree of deformation. This flat and weakly prolate deformed structure can be seen in a wide range of the neutron number, $N = 60$–72. These trends of the Skyrme HF+BCS energy surfaces are nicely reproduced by the IBM energy surfaces. Unlike some Ru isotopes in Fig. 4.2, the HF+BCS energy surface for Pd isotopes is rather simple overall to be reproduced well by the IBM energy surface.

4.2 Axially to γ-Unstable Deformed Nuclei

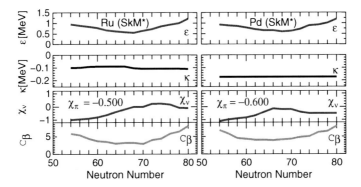

Fig. 4.4 Evolution of the IBM parameters for Ru and Pd isotopes studied with SkM* force. χ_π is kept constant as -0.500 and -0.600 for Ru and Pd, respectively. The figure has been taken from Ref. [11]

Derived IBM Parameters

We show in Fig. 4.4 the evolution of the derived IBM parameters. The overall tendencies of ε, κ and C_β are common for both Ru and Pd isotopes, where almost parabolic systematics with N with respect to the midshell is seen in ε and C_β, although there are quantitative differences to a certain extent. The χ_π value is kept constant as -0.5 and -0.6 for Ru and Pd isotopes, respectively. For Ru isotopes, χ_ν becomes larger with the neutron number N, but slightly decreases for $N \geq 74$, while, for Pd isotopes, it increases from $N = 54$ to 66, around which it becomes constant, and begins to decrease from $N = 70$. For $N \geq 66$, the magnitude of $\chi_\pi + \chi_\nu$ is slightly larger with a negative sign in the Pd isotopes than in the Ru isotopes, reflecting that the energy surface of the former is somewhat steeper in the γ direction than that of the latter, while both are similarly flat in β direction. The variation of C_β reflects the gradual change of $\beta_{\min}$ at which minimum occurs. In the earlier phenomenological work within IBM-2 [13], the parameter χ_ν increases monotonically and χ_π is opposite in sign to the present parameter values for both Ru and Pd isotopes. Other parameters used in [13] are generally consistent with the present ones.

Spectra

Figure 4.5 exhibits the evolution of low-lying spectra as functions of the neutron number N for Ru and Pd isotopes.

For both Ru and Pd isotopes, the calculated spectra in the vicinity of the shell closure $N = 50$ look like those of a spherical vibrator, where 4_1^+, 0_2^+ and 2_2^+ form the triplet, which is characteristic of the U(5) limit. This level structure is commonly found to continue from $N = 54$ to around $N = 62$. For $N = 64$–70, each calculated level comes down with the increase of N to show as a whole the O(6)-like level

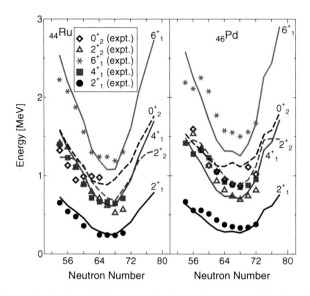

Fig. 4.5 Level evolution in Ru and Pd isotopes. *Symbols* and *curves* stand for the calculated results and the experimental data [12], respectively, as indicated in the figure. The figure has been taken from Ref. [11]

scheme. Of the triplet 4_1^+, 0_2^+ and 2_2^+, only the 0_2^+ state is not lowered and remains close to the 6_1^+ state. This is characteristic of the γ-unstable O(6) nuclei. Thus $N = 62$ nuclei are supposed to be critical points of the U(5)-O(6) transition. For the Ru isotopes, the behavior of the calculated 0_2^+ level is in good agreement with experimental one, while the experimental 4_1^+, 0_2^+ and 2_2^+ levels remain degenerated all the way, being characteristic of the vibrational level pattern. However, the calculated 0_2^+ energy of Pd isotopes is somewhat larger (or more rotational-like) than the experiment.

The HF+BCS energy surfaces for $N = 64$–70 Ru nuclei are quite soft in γ, where the depth of their minima are only a few hundred keV in energy, and can be thus approximated by totally flat IBM energy surfaces. In spite of the simplification, one obtains the good agreement of the excitation spectra. As far as the low-lying states like those in Figs. 4.2 and 4.3 are concerned, exact location of the absolute minimum does not seem to matter too much, as the triaxial dynamics can be incorporated into the spectra through the configuration mixing by, or the diagonalization of, the IBM Hamiltonian. The same applies to Ba and Xe isotopes in $A \sim 130$ region to be discussed in Sect. 4.2.2. For the precise description of the quasiγ-band energies, however, some minor contribution which produces a triaxial minimum may need to be added to the boson Hamiltonian, such as the three-body or cubic term [14–18]. In fact, the role of the cubic term has been discussed in the context of the odd-even staggering in the quasi-γ band of Ru isotopes [17, 18]. While such role played by the cubic term in the quasi-γ band of non-axially symmetric nuclei is out of the focus of

4.2 Axially to γ-Unstable Deformed Nuclei

Sect. 4.2, its importance will be noted in Sect. 4.3 and Chap. 5 and will be included within the microscopic IBM-2 framework in Chap. 6.

At $N = 72$, the calculated 4_1^+, 0_2^+ and 2_2^+ levels form the triplet consistently with the experimental data, where the overall level pattern resembles the vibrational level structure. The present calculation predicts this level pattern continues for $N \geq 74$. In each isotopic chain there seems to be another critical point of the O(6)-U(5) transition around $N = 70$.

B(E2) Ratios

In Fig. 4.6, we show the evolution of the $B(E2)$ ratios R_1–R_4 for Ru and Pd isotopes as functions of N. which are the same as those considered for Sm isotopes in Sect. 2.4.1 and were defined in Eq. (2.51). The calculated R_1 ratio is close to the O(6) limit, $R_1 = \frac{10}{7}$, being consistent with the experiments. R_2 seems to be more sensitive to N, reflecting the structural change. In Ru isotopes, R_2 increases sharply from $N = 54$ to 60 and changes much less for $N = 60$–66, reflecting the sustained γ instability. From there it goes up again and has a maximum value at $N = 72$. The experimental data show the opposite dependence on N, whereas they have large error bars. A sudden drop from around $N = 72$ can be seen in the calculated result for the Ru

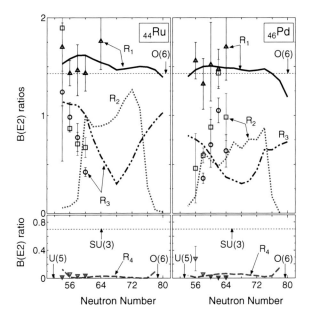

Fig. 4.6 The $B(E2)$ ratios for Ru (*left panel*) and Pd (*right panel*) isotopes as functions of N for $54 \leq N \leq 80$. *Curves* stand for the calculated $B(E2)$ ratios R_1–R_4, which are defined in Eq. (2.51). Experimental data (represented by the *symbols*) are taken from Refs. [19–26]. The figure has been taken from Ref. [11]

isotopes, approaching zero around $N = 78$. For the Pd isotopes, R_2 has a similar, but somewhat weaker dependence on N compared with the Ru isotopes, and suggests smaller values in the open-shell region. The behavior of R_3 is rather simple in the present study. For both Ru and Pd isotopes, it decreases away from the closed shells to the open-shell region consistently with the experimental data. The calculated R_4 values both for Ru and Pd isotopes are close to zero (U(5) and O(6) limits) all the way, similarly to the experimental data.

4.2.2 $Z > 50$, $50 \leq N \leq 82$ Major Shells

Energy Surface

Figure 4.7 shows the comparisons of energy surfaces for Ba isotopes between HF+BCS (indicated by HF in Fig. 4.7) calculation and the mapped IBM. The self-consistent HF+BCS energy surface for $N = 54$ indicates a weakly-deformed shape with the minimum at $\beta_{min} \sim 0.20$.

Away from the closed shells to the open-shell region, the HF+BCS energy surface becomes sharper particularly for the β direction. Accordingly, the minimum point β_{min} shifts away from the origin. Somewhat sharp prolate minima can be found from around $N = 56$ to 74, beyond which the HF+BCS energy surface becomes flat in both β and γ directions. This reflects the transition from (prolate) deformed to γ-unstable shapes. The HF+BCS energy surface suggests a nearly spherical shape with a small β_{min} value near the magic number $N = 82$. These transitions of the microscopic HF+BCS energy surface are reproduced well by the IBM energy surface.

Both the original and the mapped energy surfaces for $N = 76$ and 78 Ba nuclei have large flat areas compared with other Ba nuclei. In the present case, the energy surface of ^{134}Ba ($N = 78$) is flatter than that of ^{132}Ba ($N = 76$): ^{134}Ba seems to be close to E(5) critical-point symmetry [4], while ^{132}Ba is closer to O(6) limit of IBM. Indeed, while the $R_{4/2}$ value of the E(5) model is 2.19, the experimental value for ^{134}Ba is $R_{4/2} = 2.31$, which agrees better with the calculated result, 2.50.

We also show in Fig. 4.8 the energy surfaces for Xe isotopes, which exhibit similar systematics to, but are softer and less deformed than, Ba isotopes. The HF+BCS energy surface tends to show a sharp prolate minimum for the open-shell nuclei, and becomes flat for $N = 76$ and 78. $N = 80$ nucleus is nearly spherical, which is slightly deformed. The IBM energy surfaces reproduce all these transitions well.

Derived IBM Parameters

We show in Fig. 4.9 the evolution of the derived IBM parameters with the neutron number N for Ba and Xe isotopes, both of which show similar tendencies. While χ_ν increases with N, the quantity $\chi_\pi + \chi_\nu$ is negative all the way and becomes almost zero

4.2 Axially to γ-Unstable Deformed Nuclei

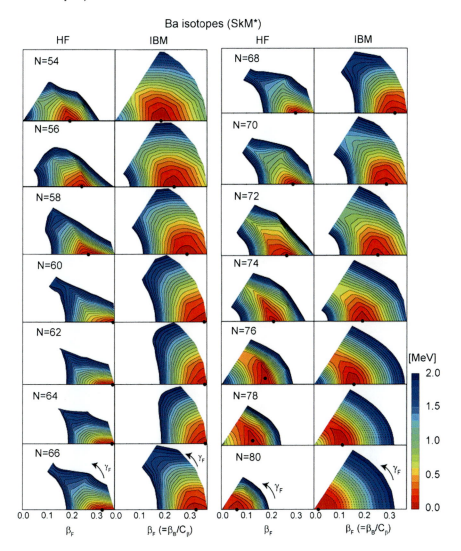

Fig. 4.7 Same as Fig. 4.2, but for Ba isotopes

for $N \geq 76$, where the γ softness appears in the energy surface. χ_π is kept constant as $\chi_\pi = -0.5$ and -0.6 for Ba and Xe isotopes, respectively. The parameters ε, κ and C_β exhibit parabolic tendency in Fig. 4.9, being maximal around the middle of the major shell at which the energy surface shows the largest deformation. The ε and the κ values for Xe isotopes are generally larger than those for Ba isotopes. The overall behaviors of the derived parameters in Fig. 4.9 are consistent with existing phenomenological IBM-2 studies [27], while the magnitude of κ in the present case is much larger than the one used there.

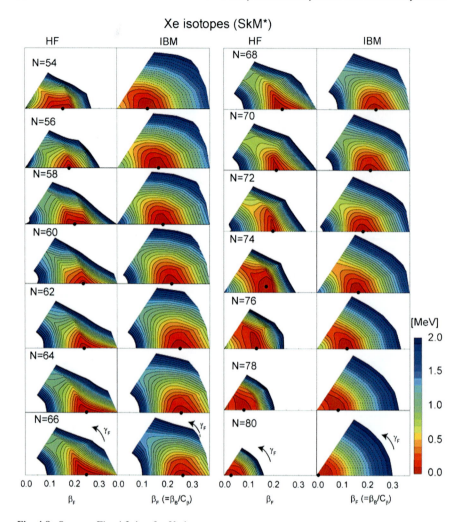

Fig. 4.8 Same as Fig. 4.2, but for Xe isotopes

Spectra

We show in the left panel of Fig. 4.10 the low-lying spectra for Ba isotopes as functions of N. The calculated yrast levels are particularly in good agreement with the experimental ones. From $N = 54$ to 58, the present calculation suggests that the side-band levels, 0_2^+ and 2_2^+, deviate from 4_1^+ level, exhibiting the transition from the nearly spherical to deformed shapes. When approaching the middle of the major shell, the calculated yrast levels decrease with N consistently with the experimental data, while the present 0_2^+ level shows an opposite dependence on N to the experiments. In the open-shell region, the calculated levels resembles rotational spectra. Indeed

4.2 Axially to γ-Unstable Deformed Nuclei

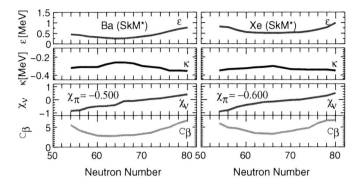

Fig. 4.9 Evolution of the derived IBM parameters for Ba and Xe isotopes with N. χ_π is kept constant as $\chi_\pi = -0.5$ and -0.6 for Ba and Xe isotopes, respectively. The figure has been taken from Ref. [11]

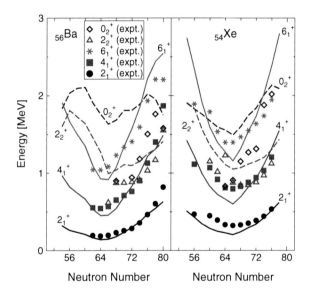

Fig. 4.10 Same as Fig. 4.5, but for Ba and Xe isotopes

the $R_{4/2}$ values of $N = 62$, 64 and 66 in the present calculation are 3.23, 3.23 and 3.14, respectively, being close to the SU(3) limit ($R_{4/2} = 3.33$), while those of the experiments are 2.86, 2.92 and 2.90, respectively. There are some deviations of the side-band levels for lighter Ba isotopes. In the present study, however, one cannot always obtain much information about the side-band structures only by referring to the unprojected HF+BCS energy surface. The improvement of the side-band levels should be an interesting future subject.

Approaching the shell closure $N = 82$, each level energy becomes larger with N and one sees the level structure of γ-unstable nuclei at $N = 76$ and 78 similarly to the experimental data. In fact, the 4_1^+ state is pushed up to be relatively close to the 2_2^+ and 0_2^+ states. This is a characteristic feature of the γ-unstable nuclei. At $N = 80$, 4_1^+, 0_2^+ and 2_2^+ states lie close to each other, which is characteristic of the spherical vibrator. In addition, the calculated $R_{4/2}$ values for $N = 76, 78$ and 80 Ba nuclei are 2.58, 2.50 and 2.36, respectively, where the first two are close to the O(6) ($R_{4/2} = 2.50$) limit and the last one U(5) limit ($R_{4/2} = 2.00$). Experimental $R_{4/2}$ values for $N = 76, 78$, and 80 Ba nuclei are 2.58, 2.31 and 2.28, respectively, which are fairly close to the present calculations.

The right panel of Fig. 4.10 shows the excitation spectra for Xe isotopes. The experimental tendencies are reproduced by the present calculations fairly well. As already indicated by the self-consistent HF+BCS energy surfaces in Fig. 4.7, the calculated excitation spectra for lighter Xe nuclei are somewhat more rotational-like, compared with the experiments. The properties of γ-unstable structures for $N = 76$ and 78 nuclei are well reproduced. Indeed, the calculated and the experimental $R_{4/2}$ values for $N = 76$ (78) Xe nucleus are 2.48 (2.54) and 2.25 (2.94), respectively, which are fairly consistent.

The HF+BCS energy surfaces for ^{132,134}Ba and ^{128}Xe have triaxial minima in between $\gamma = 0°$ and $60°$, while the IBM energy surfaces do not reproduce them. In the present study, a triaxial minimum in the HF+BCS energy surface is approximated by a flat IBM energy surface by putting $\chi_\pi + \chi_\nu \simeq 0$ in Eq. (2.46). As far as the ground-state band energies and the bandhead of the quasi-γ band 2_2^+ excitation energies are concerned, the mapped IBM gives good agreement with the data for the γ-soft nuclei. Concerning the issue of triaxiality, we here note the equivalence ansatz that the angular-momentum projection of the IBM wave function for γ-unstable system can generate similar level pattern of the rigid triaxiality [28]. The issue of whether γ-soft nucleus is γ-rigid or unstable will be addressed in Chap. 6.

$B(E2)$ Ratios

We show in Fig. 4.11 the $B(E2)$ ratios for Ba and Xe isotopes as functions of N. The calculated R_1 value does not show a strong dependence on N for both Ba and Xe isotopes, while it increases toward the middle of the major shell, being in the vicinity of the O(6) limit, $R_1 = \frac{10}{7}$. R_2 changes rather significantly. For $N = 54$–64 the calculated R_2 value is in the vicinity of the SU(3) limit, $R_2 = 0$, and becomes larger toward the shell closure $N = 82$, taking the maximal value close to O(6) limit, $R_2 = \frac{10}{7}$ at around $N = 76$ or 78. In the open-shell region, the present value of R_2 for Ba isotopes is closer to zero than Xe isotopes, where the former indicates more rotational feature of SU(3) limit than the latter. For Xe isotopes, the calculated R_2 shows a similar trend to the available data. At $N = 78$, the present R_2 value for Ba isotopes is closer to the O(6) limit than the Xe isotopes. The R_3 value does not change too much and is close to zero for open-shell nuclei. Its sharp increase for $N \geq 78$ may indicate the transition from deformed to γ-soft or to a nearly spherical shape. R_3 of Ba isotopes is smaller than that of Xe isotopes all the way, which suggests stronger deformation. R_4 increases monotonically from around the shell closures to

4.2 Axially to γ-Unstable Deformed Nuclei

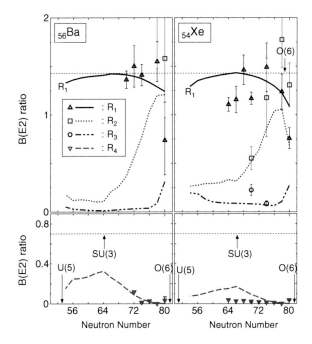

Fig. 4.11 Same as Fig. 4.6, but for Ba and Xe isotopes. Data are taken from Refs. [29–40]

the middle of the major shell at which it becomes maximal. R_4 of Ba isotopes is generally larger than that of Xe isotopes. The calculated values for both isotopes are fairly consistent with the experiment for $N = 72$–80.

Comparison with the E(5) Level Scheme

We discuss a particular nucleus, ^{134}Ba, which has been recognized as a manifestation of E(5) symmetry [6]. Figure 4.12 shows detailed level schemes of (a) the experimental data [12, 34–40], (b) the calculated result for ^{134}Ba and (c) E(5) model. Note that the 2_1^+ energy of E(5) is adjusted to 605 keV, which is the experimental 2_1^+ energy for ^{134}Ba.

In the E(5) model, a schematic potential is assumed in addition to the infinite-N limit [4]. This is apparently not the case with actual nuclei, which results in the deviations of the calculated and the experimental excitation levels from E(5) ones as seen from Fig. 4.12. Indeed, the 6_1^+, 4_2^+, 3_1^+ and 0_3^+ levels are degenerate in the E(5) model, while the overall patterns of the experimental and the calculated level schemes for ^{134}Ba seem to resemble O(6) rather than E(5).

The present values of the B(E2) ratios for $4_1^+ \to 2_1^+$ and $2_2^+ \to 2_1^+$ transitions are smaller than the experimental data, while the 0_2^+ level and the B(E2) ratios for $0_2^+ \to 2_1^+$ and $0_2^+ \to 2_2^+$ transitions agree with the experiments nicely. From the

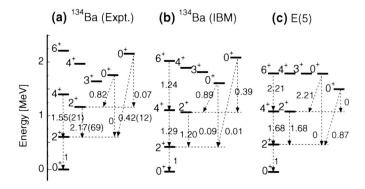

Fig. 4.12 Level schemes and $B(E2)$ ratios for (**a**) the experimental [12, 34–40] and (**b**) the calculated (IBM from SkM*) results for ^{134}Ba, and (**c**) E(5) model. The 2_1^+ energy of E(5) model is set equal to the experimental one, $E(2_1^+) = 605$ keV. The figure is taken from Ref. [11]

trends of $0_2^+ \rightarrow 2_1^+$ and $0_2^+ \rightarrow 2_2^+$ transitions, the 0_2^+ state in the present calculation may be related to the third 0^+ state of the E(5) level scheme in Fig. 4.12c, which, in terms of the ξ and the τ quantum numbers of E(5) model [4], corresponds to the $0_{\xi=1, \tau=3}$ state.

4.2.3 E(5) Symmetry in Exotic Nuclei

Having done the reasonable comparisons with experiments for various medium-heavy nuclei, we describe heavy exotic nuclei, W and Os isotopes with $N > 126$. They are chosen because no systematic theoretical work has been done.

Mapped Energy Surfaces and Derived IBM Parameters

The self-consistent HF+BCS and the mapped IBM energy surfaces for W and Os isotopes are compared in Fig. 4.13 for $N = 130$–140. The HF+BCS energy surfaces for $N = 130$ nuclei, ^{204}W and ^{206}Os, have the minima at $\beta \sim 0$, being similar to the harmonic oscillator potential characteristic of the vibrational or U(5) limit. The corresponding IBM energy surfaces generally look somewhat flatter, but the overall patterns are almost the same as the HF+BCS ones. In both W and Os isotopes, the location of the energy minimum shifts gradually to $\beta \neq 0$ with N and the energy surface becomes steeper in the γ direction, while the flat area becomes larger. This large flat area, characteristic of E(5) symmetry, continues from $N = 132$ to 136, while the change of the energy surface for Os isotopes looks moderate in comparison to the W isotopes. For the $N = 134$ nuclei, ^{208}W and ^{210}Os, for instance, which are located on the way to the shape transition, the energy surfaces exhibit a typical O(6)-E(5)

4.2 Axially to γ-Unstable Deformed Nuclei

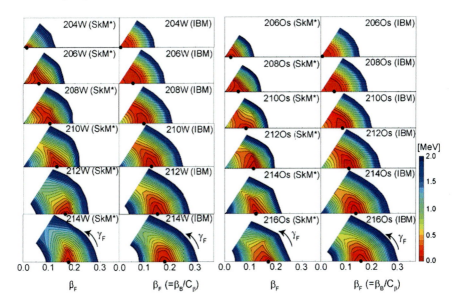

Fig. 4.13 Same as Fig. 4.2, but for W and Os isotopes

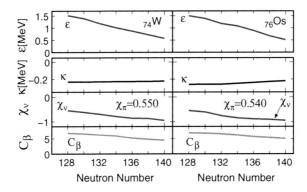

Fig. 4.14 Evolution of the IBM parameters with N for W and Os isotopes. χ_π is kept constant as $\chi_\pi = 0.55$ and 0.54 for W and Os isotopes, respectively. The figure has been taken from Ref. [11]

structure similarly to the ^{132,134}Ba nuclei. Indeed, the predicted $R_{4/2}$ values for ^{208}W and ^{210}Os are 2.49 and 2.45, respectively, both of which are close to the E(5) value (=2.19) and to the experimental $R_{4/2}$ value of ^{134}Ba (=2.31). For $N \geq 140$, the HF+BCS energy surface predicts stronger deformation.

We show in Fig. 4.14 the evolution of the derived IBM parameters for W and Os isotopes. Of particular interest is that χ_π and χ_ν have opposite signs with sizable magnitudes for $N \geq 130$. In the IBM-2, this is the origin of the O(6)-E(5) pattern [41–43]. Each parameter does not change too much with N since there is no drastic change of the energy surface.

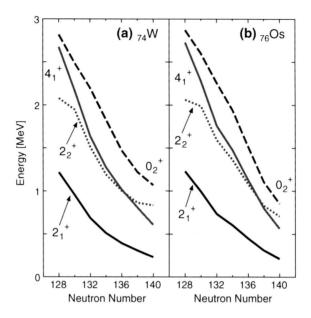

Fig. 4.15 Calculated excitation levels of (**a**) W and (**b**) Os isotopes as functions of N. The figure has been taken from Ref. [11]

Spectra and $B(E2)$ Trends

In Fig. 4.15 the level evolution for W and Os isotopes is shown. It is of considerable interest that the magnitude of deformation, represented by the decrease of the 2_1^+ excitation energy, becomes larger with N, while the γ-unstable E(5)-O(6) level pattern is maintained all the way. Such sustained E(5)-O(6) level structure has never been seen in stable nuclei, and may become one of the characteristic features of exotic nuclei with considerable neutron excess. While we assume in the present study that the proton and the neutron systems move in phase, these restrictions could be relaxed. The mechanism which causes such an unexpectedly large region of the E(5) pattern would be also studied in the future. For $N \geq 136$ or 138, the calculated 4_1^+ level continues to decrease with N, while the 2_2^+ state gradually increases. This indicates the structural evolution from γ-soft to axially symmetric deformed nuclei. While these tendencies can be found commonly in W and Os isotopes, the transition in Os isotopes occurs moderately compared with W isotopes.

We show in Fig. 4.16 the predicted $B(E2)$ ratios. While R_1 is almost constant, being close to the O(6) limit ($=\frac{10}{7}$), R_2 becomes larger with N and becomes maximal at around $N = 132$ for W isotopes and around $N = 132$ or 134 for Os isotopes. The R_2 value looks closer to that of the O(6) limit and that of E(5) model ($=1.67$) for $N = 132$–136, while for Os isotopes, R_2 changes with N less significantly than for W isotopes. This moderate change of R_2 may be also a characteristic feature of the sustained γ softness which can be found in the energy surface and in the

4.2 Axially to γ-Unstable Deformed Nuclei

Fig. 4.16 Predicted $B(E2)$ ratios for W (*solid curves*) and Os (*dashed curves*) isotopes as functions of the neutron number N. The ratios R_1–R_4 are defined in Eq. (2.51). The Skyrme functional SkM* is used. The figure has been taken from Ref. [11]

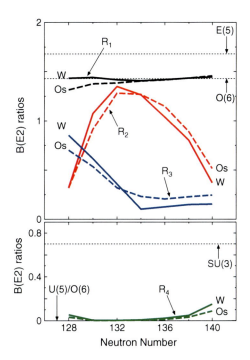

excitation spectra. R_3 gradually decreases with N similarly to the Ru isotopes, which suggests structural change. R_4 remains almost zero, being much below the SU(3) limit, $R_4 = \frac{7}{10}$.

Comparison with the E(5) Level Scheme

For the sake of completeness, we show in Fig. 4.17 the detailed level schemes, focusing on the $N = 134$ nuclei, ^{208}W and ^{210}Os, which may be candidates for E(5) critical points. The calculated results of (b) ^{208}W and (c) ^{210}Os are compared with (a) the experimental level scheme for ^{134}Ba nucleus [12, 34–40]. The calculated $B(E2)$ is generally smaller than the E(5) one of Fig. 4.12c. For both ^{208}W and ^{210}Os, the calculated $B(E2)$ ratios for the transitions from the yrast to the side-band levels show quite similar trends to the experiments for ^{134}Ba, except for the selection rule for the $0_3^+ \rightarrow 2_1^+$ and the $0_3^+ \rightarrow 2_2^+$ E2 transitions. The calculated $B(E2)$ ratios for $0_2^+ \rightarrow 2_2^+$ transition are 1.55 and 1.09 for ^{208}W and for ^{210}Os, respectively, both of which are closer to the E(5) value (=2.17) than the calculated value for ^{134}Ba (=0.89) in Fig. 4.12b and the experimental data for ^{134}Ba (=0.82). Moreover, particularly for ^{208}W, levels of $6_1^+, 4_2^+, 3_1^+$ and 0_2^+ states are to a greater extent degenerated than those of ^{134}Ba, which is rather O(6)-like. This may also be an evidence for the richness of the $N > 126$ region mass region in that there are many examples of E(5)-like nuclei.

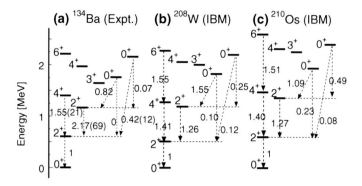

Fig. 4.17 Level schemes and $B(E2)$ ratios. **a** Experimental data for ^{134}Ba [12, 34–40], calculated results for **b** ^{208}W and **c** ^{210}Os. Skyrme functional of SkM* is used. The figure has been taken from Ref. [11]

4.3 Prolate-Oblate Shape Dynamics

4.3.1 IBM from Gogny D1S

In this section, we present spectroscopic calculations for the Pt isotopic chain (i.e., for the even-even isotopes $^{172-200}$Pt) in terms of an IBM Hamiltonian determined microscopically by mapping the energy surface obtained in the framework of the (constrained) Hartree-Fock-Bogoliubov (HFB) approximation [44–46] based on the parametrization D1S [47] of the Gogny-EDF [48, 49]. Quite recently, the structural evolution in Pt isotopes, including the role of triaxiality (i.e., the γ degree of freedom), has been studied by Rodríguez-Guzmán et al. [50]. In addition to the (standard) Gogny-D1S EDF, the new incarnations D1N [51] and D1M [52] of the Gogny-EDF have also been included in the mean-field analysis of Ref. [50]. The considered range of neutron numbers included prolate, triaxial, oblate and spherical shapes and served for a detailed comparison of the (mean-field) predictions of the new parameter sets D1N and D1M against the standard parametrization D1S. It has been shown that, regardless of the particular version of the Gogny-EDF employed, the prolate-to-oblate shape/phase transition occurs quite smoothly with the γ-softness playing an important role. It is therefore very interesting to study how the systematics of the HFB energy surfaces discussed in Ref. [50] is reflected in the isotopic evolution of the corresponding low-lying quadrupole collective states and how accurately such states can be reproduced by a mapped IBM Hamiltonian [11, 53]. Let us stress that our main goal of this section is to study the performance of a fermion-to-boson mapping procedure [11, 53] based on the Gogny-EDF. For this reason, as a first step, we will restrict ourselves to a mapping in terms of the parametrization Gogny-D1S already considered as global and able to describe reasonably well low-energy experimental data all over the nuclear chart (see, for example, Refs. [46, 54] and references therein).

4.3 Prolate-Oblate Shape Dynamics

From the theoretical perspective, the Pt and neighboring isotopic chains have been extensively studied in terms of both IBM and mean-field-based approaches. There is much experimental evidence [55, 56] revealing existences of γ-unstable O(6) nuclei in Pt isotopes. The IBM-2 has been used in a phenomenological way for the spectroscopy of Pt, Os and W isotopes [57, 58]. The prolate-to-oblate transition in Pt as well as in Os and W nuclei, has been observed in the recent experiment [59], where a relatively moderate oblate-to-prolate shape/phase transition occurs in Pt as compared to Os and W nuclei. Spectroscopic calculations have been carried out for Pt isotopes in the framework of the five-dimensional collective Hamiltonian, derived from the pairing-plus-quadrupole model [60]. Evidence for γ vibrations and shape evolution in $^{184-190}$Hg has been considered in Ref. [61], where a five-dimensional collective Hamiltonian was built with the help of constrained Gogny-D1S HFB calculations. On the other hand, systematic mean-field studies of the evolution of the ground state shapes in Pt and the neighboring Yb, Hf, W and Os nuclei have been carried out with non-relativistic Skyrme [44] and Gogny [46, 50] EDFs, as well as within the framework of the relativistic mean-field (RMF) approximation [62]. One should also keep in mind, that Pt, Pb and Hg nuclei belong to a region of the nuclear chart, around the proton shell closure $Z = 82$, characterized by a pronounced competition between low-lying configurations corresponding to different intrinsic deformations [63] and therefore, a detailed description of the very rich structural evolution in these nuclei requires the inclusion of correlations beyond the static mean-field picture [64–66] accounting for both symmetry restoration and configuration mixing. The role of configuration mixing in this region has also been considered in phenomenological IBM studies [67–69].

Description of the Model

In order to compute the Gogny-HFB energy surfaces, which are our starting point, we have used the (constrained) HFB method together with the parametrization D1S of the Gogny-EDF. The solution of the HFB equations, leading to the set of vacua $|\Phi_{HFB}(\beta, \gamma)\rangle$, is based on the equivalence of the HFB with a minimization problem that is solved using the gradient method [46, 70]. In agreement with the fitting protocol of the force, the kinetic energy of the center of mass motion has been subtracted from the Routhian to be minimized in order to ensure that the center of mass is kept at rest. The exchange Coulomb energy is considered in the Slater approximation and we neglect the contribution of the Coulomb interaction to the pairing field. The HFB quasiparticle operators are expanded in a Harmonic Oscillator (HO) basis containing enough number of shells (i.e., $N_{shell} = 13$ major shells) to grant convergence for all values of the mass quadrupole operators and for all the nuclei studied. We constrain the average values of the mass quadrupole operators $\hat{Q}_{20} = \frac{1}{2}\left(2z^2 - x^2 - y^2\right)$ and $\hat{Q}_{22} = \frac{\sqrt{3}}{2}\left(x^2 - y^2\right)$ to the desired deformation values Q_{20} and Q_{22} defined as

$$Q_{20} = \langle \Phi_{\text{HFB}} | \hat{Q}_{20} | \Phi_{\text{HFB}} \rangle \tag{4.2}$$

and

$$Q_{22} = \langle \Phi_{\text{HFB}} | \hat{Q}_{22} | \Phi_{\text{HFB}} \rangle. \tag{4.3}$$

In Ref. [50], the $Q - \gamma$ energy contour plots with

$$Q = \sqrt{Q_{20}^2 + Q_{22}^2} \tag{4.4}$$

and

$$\tan \gamma = \frac{Q_{22}}{Q_{20}} \tag{4.5}$$

have been used to study the (mean-field) evolution of the ground state shapes in Pt nuclei. Alternatively, one could also consider the $\beta - \gamma$ representation in which the quadrupole deformation parameter β is written [46] in terms of Q [Eq. (4.4)] as

$$\beta = \sqrt{\frac{4\pi}{5}} \frac{Q}{A \langle r^2 \rangle} \tag{4.6}$$

where $\langle r^2 \rangle$ represents the mean squared radius evaluated with the corresponding HFB state $|\Phi_{\text{HFB}}\rangle$.

The set of constrained HFB calculations described above, provides one with the Gogny-D1S $\beta - \gamma$ energy surface (i.e., the total HFB energies $E_{\text{HFB}}(\beta, \gamma)$ [71]) required for the subsequent mapping procedure, for which the basic IBM-2 Hamiltonian of Eq. (2.41). This type of the Hamiltonian used in this chapter embodies only single configuration, i.e., the one that does not include the intruder configuration, which arises from the particle-hole excitation across the proton shell closure. This means that, in the present work, we do not touch on the shape-coexistence phenomena. The validity of the use of the Hamiltonian of single configuration will be examined in Sect. 4.3.1. Here, the ^{132}Sn and the ^{208}Pb doubly-magic nuclei are assumed to be inert cores, while neutron boson number is changed according to the usual boson counting rule. This also applies to the analyses in Sects. 4.3.2 and 4.3.3.

Gogny-D1S and Mapped IBM Energy Surfaces

The IBM energy surfaces obtained for the nuclei $^{180-198}$Pt are shown in Fig. 4.18 as a representative sample. The IBM parameters ε, κ, χ_π, χ_ν and C_β, to be discussed later on, have been obtained by mapping the corresponding Gogny-D1S energy surfaces are also shown as references in Fig. 4.18, which are identical to those presented in Fig. 2 of Ref. [50] along the lines described before.

4.3 Prolate-Oblate Shape Dynamics

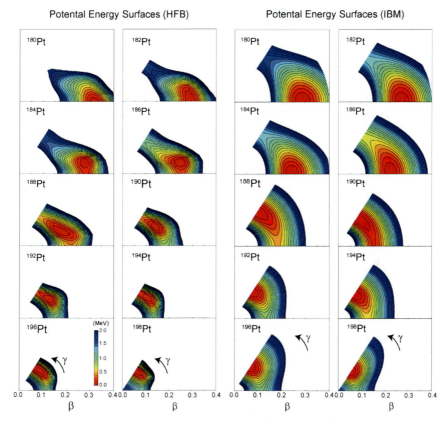

Fig. 4.18 The HFB and the IBM total energy landscapes in $\beta\gamma$ plane for the nuclei $^{180-198}$Pt. Here, $\gamma = \gamma_B$ and $\beta = \beta_B/C_\beta$. The energy surfaces are shown within $0.00 \leq \beta \leq 0.40$ and $0° \leq \gamma \leq 60°$ up to 2 MeV excitation from the minimum. Contour spacing is 100 keV. For details, see the main text. The figures are taken from Ref. [72]

The IBM energy surfaces from ^{180}Pt to ^{186}Pt display a prolate deformed minimum and an oblate deformed saddle point. The prolate minimum becomes softer in γ but steeper in β direction as the number of neutrons increases. This is, roughly speaking, consistent with the topologies of the HFB energy surfaces, where the minima are located a bit off but quite nearby the line $\gamma = 0°$.

The IBM energy surfaces for both ^{188,190}Pt are γ soft, having the minimum on the oblate side. These nuclei are supposed to be close to the critical point of the prolate-to-oblate shape transition. The corresponding HFB energy surfaces display shallow triaxial minima with $\gamma \sim 30°$ and are also soft along the γ direction. The IBM Hamiltonian considered in the present study does not provide a triaxial minimum, but either prolate or oblate minimum, as can be seen from Eq. (2.46). The γ-softness can be simulated by choosing the parameters χ_π and χ_ν so that their sum becomes nearly equal to zero. This is reasonable when a triaxial minimum is not deep enough like the present case, where the triaxial minimum point in the HFB energy surface differs by at most several hundred keV in energy from either prolate or oblate saddle point. However, the topology of the mapped IBM energy surface is then somewhat sensitive to the values of the parameters χ_π and χ_ν, which occasionally results in a quantitative difference in the location of the minimum in the IBM energy surface from that of the HFB energy surface. In fact, and contrary to what happens with the HFB energy surfaces, the IBM energy surface of ^{190}Pt is softer in γ than that of ^{188}Pt. One should then expect a certain deviation of the resultant IBM spectra from the experimental ones, which can be partly attributed to the small difference already mentioned.

In Fig. 4.18, isotopes from ^{192}Pt to ^{198}Pt exhibit oblate deformation. The locations of their energy minima and their curvatures in both β and γ directions agree well with the ones of the Gogny-D1S energy surfaces. These isotopes become steeper in the γ direction and shallower in the β direction as the number of neutrons increases. Their energy minima approach the origin more rapidly than the lighter Pt nuclei shown Fig. 4.18. This evolution reflects the transition from oblate deformed ground states to a spherical vibrator as one approaches the neutron shell closure $N = 126$.

Derived IBM Parameters

The IBM parameters ε, κ, $\chi_{\pi,\nu}$ and C_β derived for the nuclei $^{172-200}$Pt from the mapping procedure described are depicted in Fig. 4.19a–d as functions of the mass number A.

Figure 4.19a shows the parameter ε gradually decreases toward mid shell in accordance with the growth of the deformation. This trend reflects the structural evolution from nearly spherical to more deformed shapes and is consistent with previous results for other isotopic chains [11]. In Fig. 4.19b, the derived κ parameter is almost constant and somewhat larger in comparison with the phenomenological value [58], which is the consequence of the sharp potential valleys observed in the Gogny-D1S energy surfaces.

On the other hand, in Fig. 4.19c the proton parameter χ_π is almost constant while the neutron parameter χ_ν changes significantly. The systematic behavior of the present χ_ν value is consistent with the phenomenological one [58], while there is quantitative difference between the former and the latter. The magnitude of the sum $\chi_\pi + \chi_\nu$ as well as its sign depend on how sharp the HFB energy surface is in the γ direction and on whether the nucleus is prolate (negative sum) or oblate (positive sum) deformed, respectively. Therefore, as χ_π does not change much, the role of γ instability can be seen clearly from the systematics of χ_ν. For the isotopes $^{172-180}$Pt the energy surface exhibits prolate deformation and the sum $\chi_\pi + \chi_\nu$ has negative sign. The average of the derived χ_π and χ_ν values is nearly equal to zero for the nuclei $^{182-194}$Pt. This is a consequence of the γ softness in the corresponding HFB energy surfaces. On the other hand, the sum $\chi_\pi + \chi_\nu$ becomes larger with positive sign as we approach the neutron shell closure $N = 126$ reflecting the appearance of

4.3 Prolate-Oblate Shape Dynamics

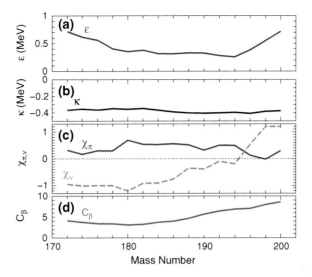

Fig. 4.19 The derived IBM parameters $\varepsilon, \kappa, \chi_\pi, \chi_\nu$ and C_β as functions of the mass number A. The meanings of these parameters were presented in Eqs. (2.41) and (2.47). For the wavelet analysis the Morlet function is used [11]. The figure is taken from Ref. [72]

weakly deformed oblate structures in the corresponding energy surfaces. In recent calculations with IBM-1 [68, 69], the sign of χ parameter (equivalent to $(\chi_\pi + \chi_\nu)/2$ in IBM-2) is always negative for $A \leq 194$.

Figure 4.19d shows that C_β decreases gradually toward the middle of the major shell. C_β can be interpreted as the "bridge" between the geometrical deformation β [73, 74] and the IBM deformation β_B and is thus proportional to the ratio between the total and valence nucleon numbers, in a good approximation [75]. This is probably the reason for the decreasing trend observed in Fig. 4.19d, as well as in earlier studies for other isotopic chains [11, 53].

Spectroscopic Calculations

With all the parameters ε, κ, $\chi_{\pi,\nu}$ and C_β required by the IBM Hamiltonian at hand, we are now able to test the spectroscopic quality of our mapping procedure, based on the Gogny-D1S EDF, for the nuclei $^{172-200}$Pt. Therefore, in the following we will discuss our predictions concerning the properties of the low-lying spectra as well as the reduced transition probabilities B(E2). We will also consider their correspondence with the mapped energy surfaces and the derived IBM parameters. We will compare our theoretical predictions with the available experimental data taken from Brookhaven National Nuclear Data Center (NNDC) [12] and from the latest Nuclear Data Sheets [76–91]. The diagonalization of the IBM Hamiltonian is performed numerically for each nucleus using the code NPBOS [92].

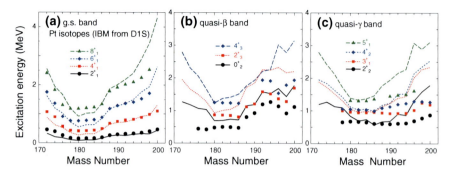

Fig. 4.20 Evolution of calculated (*curves*) and experimental (*symbols*) low-lying spectra of $^{172-200}$Pt nuclei for (**a**) ground-state, (**b**) quasi-β and (**c**) quasi-γ bands as functions of the mass number A. Experimental data are taken from Ref. [12]. The figure is taken from Ref. [72]

Here we have to note that the experimental 2_3^+ and 4_3^+ levels for mass numbers $192 \leq A \leq 200$ belong to bands different from that of the 0_2^+ level, while they are assigned to the quasi-β-band levels in Fig. 4.20b for convenience sake. Similarly, as one will see in Fig. 4.20c the experimental data for the 3_1^+ and the 4_2^+ levels in ^{198,200}Pt are assigned to the quasi-γ-band levels lying on top of the 2_2^+ energy.

Evolution of Low-Lying Spectra

Figure 4.20 displays the calculated spectra for (a) ground-state, (b) quasi-β and (c) quasi-γ bands. What is striking is the good agreement between the present calculations and the experimental data not only for ground-state but also for quasi-β and quasi-γ band energies, where overall experimental trends are reproduced fairly well in particular for the open-shell nuclei $^{180-192}$Pt.

We show in Fig. 4.20a the evolution of the 2_1^+, 4_1^+, 6_1^+ and 8_1^+ levels in the considered Pt nuclei as functions of the mass number A. The calculated energies decrease toward the middle of the major shell with the number of the valence neutrons and remain almost constant for $176 \leq A \leq 186$ nuclei. Although these tendencies are well reproduced, the rotational features are somewhat enhanced in the calculated levels for ^{180,182,184}Pt which are slightly lower in energy than the experimental ones. From both the theoretical results and the experimental data, one can also observe clear fingerprints for structural evolution with a jump between ^{186}Pt and ^{188}Pt, which can be correlated with the change of the mapped energy surfaces from prolate to oblate deformations. For $A \geq 188$ the yrast levels gradually go up as the neutron shell closure $N = 126$ is approached.

One can also find signatures for a shape/phase transition in the systematics of the quasi-β band levels shown in Fig. 4.20b. From $A = 180$ to 186, the 0_2^+ band head and the 2_3^+ level look either constant or nearly constant in both theory and experiment. The two levels are pushed up rather significantly from $A = 186$ to 188 consistently

4.3 Prolate-Oblate Shape Dynamics

with the systematics in the ground-state band and with the change of the mapped energy surfaces as functions of the neutron number N. The calculated 0_2^+ and 2_3^+ levels are higher than but still follow the experimental trends.

Coming now to the quasi-γ band levels shown in Fig. 4.20c, one can observe the remarkable agreement between theoretical and experimental spectra for $180 \leq A \leq 186$, where the 3_1^+ level lies close to the 4_2^+ level. However, the present calculation suggests this trend persists even for $188 \leq A \leq 196$, whereas the relative spacing between the experimental 3_1^+ and 4_2^+ levels for these nuclei is larger. Similar deviation occurs for 5_1^+ and 6_2^+ levels, although the latter is not exhibited in Fig. 4.20b. This means that our calculations suggest the feature characteristic of the O(6) symmetry, where the staggering occurs as 2_γ^+, $(3_\gamma^+ \ 4_\gamma^+)$, $(5_\gamma^+ \ 6_\gamma^+)$, ... etc. However, the experimental levels are lying more regularly particularly for $188 \leq A \leq 196$, and thus appear to be in between the O(6) limit and a rigid triaxial rotor where the staggering shows up as $(2_\gamma^+ \ 3_\gamma^+)$, $(4_\gamma^+ \ 5_\gamma^+)$, ... etc. [28]. Such a deviation of the γ-band structure seems to be nothing but a consequence of an algebraic nature of the IBM, and indeed has also been found in existing phenomenological IBM calculations [58]. From a phenomenological point of view, the so-called cubic (or the three-boson) interaction [14–16] has been useful for reproducing the experimental γ-band structure. The cubic term produces a shallow triaxial minimum that is seen in the Gogny-HFB energy surface, and may be introduced also in the Hamiltonian defined in Eq. (2.41). This is, however, out of focus in the current theoretical framework, because the cubic term represents an effective force whose origin remains to be investigated further.

Here, the deviations observed in the side-band levels (even in some of the ground-state band levels) for $A \geq 196$ are probably related to the larger magnitude of the parameter κ as compared with its phenomenological value [58]. Roughly speaking, when the magnitude of κ becomes larger, the moment of inertia decreases, resulting in the deviation of not only ground-state-band but also the side-band energies. The problem arises in the present case partly because, in the vicinity of the shell closure $N = 126$, the HFB energy surfaces exhibit weak oblate deformations close to the origin $\beta = 0$, as we showed in Fig. 4.18. In addition, the curvatures along the β direction around the minima are somewhat larger. These peculiar topologies of the Gogny-D1S energy surfaces make it rather difficult to determine a value of κ which gives reasonable agreement of side-band energies with the experimental ones. In this case one may interpret that the deviation is mainly due to the properties of the particular version of the Gogny-EDF considered in the present study. Another possibility is that the boson Hamiltonian used may be still simple, requiring the introduction of additional interaction terms in the boson system. Investigation along these lines is in progress and will be reported elsewhere.

We note that some lighter Pt nuclei, as well as the neighbouring Hg and Pb isotopes, are often revealed to exhibit coexistence of prolate and oblate shapes. Some existing IBM studies consider a configuration mixing, i.e., particle-hole excitation across the proton $Z = 82$ shell, leading to enlarged model space consisting of a so-called regular/single configuration (with N_B bosons) and an intruder, deformed $2p$-$2h$ excitation configuration (with $N_B + 2$ bosons) [67]. Along this line, there have been debates over which is adequate for the description of low-lying structure of lighter Pt isotopes, a single- or a configuration-mixing IBM framework [68, 69]. A simple IBM-1 Hamiltonian was applied for Pt nuclei in Ref. [68], leading to good agreement with the experimental data without a need for an intruder configuration. Moreover, in Ref. [69], both single (within IBM-1 framework) and configuration-mixing models were shown to give almost equivalent results for Pt isotopes being consistent with the experimental data, as long as the excitation energy is rather low (up to $E_x \sim 1.5$ MeV). In the present calculation, similarly to those in Refs. [68, 69], the agreement between theoretical and experimental spectra, shown in Fig. 4.20, is reasonably good without introducing an intruder configuration. Once the intruder configuration is taken into account in the present framework, the number of free parameters to be fixed would largely increase. On the other hand, it is rather hard to identify a clear shape coexistence for the considered Pt nuclei in a microscopic level, as an HFB-energy surface in Fig. 4.18, indeed, does not exhibit any isolated local minimum other than the global one. Nevertheless, to what extent our result is changed by the configuration mixing, as well as a comparison between a single and a configuration-mixing calculations, is an interesting future issue which will be investigated further.

Systematics of $B(E2)$ Ratios

Next we discuss overall systematic trends of the $B(E2)$ ratios R_1–R_4, which were studied already in Sect. 2.4.1 for Sm, Sect. 4.2.1 for Ru and Pd, and Sect. 4.2.2 for Ba and Xe, and which were defined in Eq. (2.51). The results are shown in Fig. 4.21 as functions of the mass number A. Note that the values of each dynamical symmetry limit [93, 94], to be shown below, are those with infinite boson number.

The ratio R_1 in Fig. 4.21a is nearly constant all the way, being much below the U(5) limit of IBM ($R_1 = 2$), and is rather close to $R_1 = 10/7$, which is the O(6) and SU(3) limit of IBM. Thus, R_1 is not a sensitive observable to distinguish between axially symmetric and γ-soft nuclei. This is reasonable because the structural evolution between axially symmetric deformed and the γ-unstable shapes is shown to take place quite smoothly from the systematics of the mapped energy surfaces (in Fig. 4.18) and the derived IBM parameters (in Fig. 4.19). The flat behavior of R_1 value for Pt isotopes in Fig. 4.21a differs from the one found e.g., in Sm isotopes [95]. There, a sharp decrease of R_1 value can be seen in the line of U(5)-SU(3) shape/phase transition.

One can see also in Fig. 4.21a that, in contrast to the flat systematics of the R_1 value with respect to the mass number A, the ratio R_2 changes significantly and is relatively large for $^{186-196}$Pt nuclei, being close to $\frac{10}{7}$ (O(6) limit). This is consistent with the softness of the energy surfaces for these nuclei. Therefore, the quantity R_2 is quite sensitive to the shape evolution encountered in the energy surfaces and can be thus considered as the best signature for γ-softness among R_1–R_4. There are not much available data overall, but the experimental R_2 value is also relatively large around ^{192}Pt. For the nuclei $^{176-184}$Pt, the theoretical R_2 value is close to zero (the SU(3) limit) and slightly goes up from $A = 174$ to 172, proba-

4.3 Prolate-Oblate Shape Dynamics

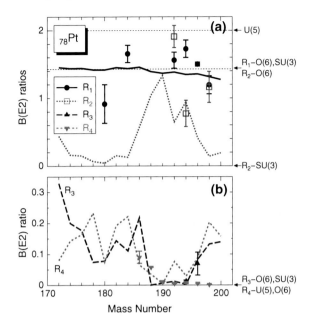

Fig. 4.21 The $B(E2)$ ratios R_1, R_2, R_3 and R_4 (defined in Eq. (2.5)) for relevant low-lying states of $^{172-200}$Pt nuclei as functions of the mass number A. Experimental data are taken from Ref. [58, 76–91]. For more details, see the main text. The figure is taken from Ref. [72]

bly approaching the U(5) vibrational limit ($R_2 = 2$) in the vicinity of the neutron shell closure $N = 82$.

Unlike the R_2 case, the calculated ratio R_3, shown in Fig. 4.21b, does not change much with mass number A and is close to zero (O(6) and SU(3) limits of R_3) for $^{188-196}$Pt. From $A = 180$, the R_3 turns to increase as we move towards the neutron shell closure $N = 82$ and is expected to approach the U(5) limit ($R_3 = 2$). The calculated R_3 value is, however, still much smaller than the experimental value at $A = 198$. In fact, both the HFB and the mapped energy surfaces for the nucleus ^{198}Pt display a weakly deformed shape, which somewhat differs from the vibrational feature expected from the corresponding experimental levels. The present R_3 value does not exhibit a drastic change observed in shape transitions in $A \sim 130$ Ba-Xe and $A \sim 100$ Ru-Pd isotopes, where the E2 transition from the 0_2^+ state to the 2_1^+ is much enhanced [11].

Finally, the branching ratio R_4 in Fig. 4.21b also corresponds to a gradual shape transition. The present calculations suggest that the R_4 value is nearly zero (O(6) limit) in the region where the nuclei are soft and where the R_2 ratio takes large values. The calculated R_4 ratio follows the experimental trend exhibiting increase from $A = 190$ to 186, and becomes relatively larger for $A \leq 184$, where the energy surfaces show stronger prolate deformation. Consistently with the evolution of the IBM energy surfaces, the calculated R_4 values turn to approach the U(5) limit, which is also zero, for $A \leq 178$. Similarly to the R_3 case, a deviation from the vibrational character of the experimental data is found at $A = 198$.

It should be emphasized that all the results for $B(E2)$ values shown so far are quite consistent with the topologies of the energy surfaces and with the derived IBM parameter values.

Level Schemes of Selected Nuclei

As already mentioned in the introductory part of this section, one of the main goals of the present study is to test the spectroscopic quality of the mapping procedure and the underlying (universal) Gogny-D1S EDF [46, 54]. Keeping this in mind, we will now turn our attention to a more detailed comparison between our results and the available experimental data for excitation spectra and $B(E2)$ values. To this end, we select the nuclei $^{184-194}$Pt as a representative sample corresponding to the mapped energy surfaces shown in Fig. 4.18. The level schemes obtained for the nuclei $^{184-194}$Pt are compared in Fig. 4.22 with available experimental data. The theoretical $B(E2)$ values are shown also in Fig. 4.22. Note that $B(E2; 2_1^+ \to 0_1^+)$ value is normalized to the experimental one. The virtue of the present calculation is to give predicted $B(E2)$ values for those nuclei which have no enough E2 information available. This is particularly useful in the cases of ^{184}Pt, ^{186}Pt and ^{188}Pt where the calculated spectra agree well with the experiment.

For clarity, we divide the explanation of the results shown in Fig. 4.22 into the following three categories, according to the tendencies of what we found in the energy surfaces and the IBM parameters and the experimental data. The first is the prolate deformed regime represented by the nuclei ^{184,186}Pt, which exhibit a rotational character. Next, we will consider the isotopes ^{188,190}Pt which are apparently close to the critical point of the prolate-to-oblate transition observed in the mapped energy surfaces (see Fig. 4.18). Lastly, calculated and experimental results are compared for the nuclei ^{192,194}Pt which belong to the weakly oblate deformed regime. Note that, in Fig. 4.22, the energy scale is not common for all nuclei.

For ^{184,186}Pt, the present calculation reproduces overall pattern of the experimental spectra in all of the ground-state, quasi-β and quasi-γ bands fairly well. Interesting enough, the bandhead energies, particularly the quasi-β bandhead 0_2^+, are much higher than the 4_1^+ level, compared to the experimental data. This indicates that, reflecting the topologies of the energy surfaces in Fig. 4.18, the ^{184,186}Pt nuclei deviate from the γ-soft O(6) character and exhibit rather rotational features. The 0_3^+ energies are predicted to be above 4_3^+ ones in both nuclei.

On the other hand, both ^{188}Pt and ^{190}Pt, whose energy surfaces are quite flat along the γ direction in Fig. 4.18, appear to be closer to the γ-unstable O(6) limit of the IBM than the two nuclei already mentioned above. The 2_2^+ and 4_1^+ levels lie close to each other in the present study and, in the spirit of group theory, are supposed to have the same $\tau = 2$ quantum number of the O(6) dynamical symmetry [96]. Similarly, in our calculations the $6_1^+, 4_2^+, 3_1^+$ and 0_2^+ levels are almost degenerate and can be then grouped into the $\tau = 3$ multiple. Along these lines, we can observe characteristic E2

4.3 Prolate-Oblate Shape Dynamics

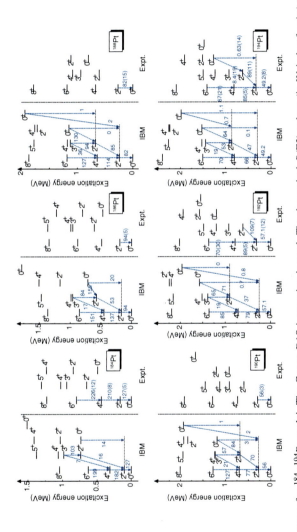

Fig. 4.22 Level schemes for $^{184-194}$Pt nuclei. The Gogny D1S interaction is used. The theoretical $B(E2)$ values (in Weisskopf units) for each nucleus are normalized to the experimental $B(E2; 2_1^+ \to 0_1^+)$ value. Note that, in ^{190}Pt, the theoretical 3_1^+ level is almost identical to but is lower by only 10 keV in energy than the 4_2^+ level, where the E2 transitions shown are those from the 3_1^+ state, not from the 4_2^+ state. For details, see the main text. The figure is taken from Ref. [72]

decay patterns that are quite consistent with the $\Delta \tau = \pm 1$ selection rule of the O(6) limit [96]. For instance, the transition from the 0_2^+ level (supposed to have $\tau = 3$) to 2_2^+ level (supposed to have $\tau = 2$) is dominant over the one to 2_1^+ (supposed to have $\tau = 1$) in both ^{188}Pt and ^{190}Pt. The trend characteristic of O(6) symmetry is clearly seen particularly in ^{190}Pt, where the sum of the parameters χ_π and χ_ν almost vanishes as seen from Fig. 4.19c. This means that the nucleus is close to the pure O(6) limit, and is consistent with the mapped energy surface in Fig. 4.18 that is nearly flat along the γ direction. Nevertheless, the structure of the corresponding experimental γ band appears to have a more triaxial nature, where the 3_1^+ and the 4_2^+ levels are apart from each other. As we have anticipated from the pattern of the energy surface in Fig. 4.18, this deviation arises partly due to the difference of the position of absolute minimum between Gogny-D1S and the corresponding IBM energy surfaces in Fig. 4.18.

For ^{192,194}Pt nuclei, the theoretical γ-band structure still looks like that of O(6) symmetry: the calculated 3_1^+ and 4_2^+ energies are almost degenerated. Similar tendency has been confirmed for ^{194}Pt nucleus in some recent IBM-1 studies [68, 69]. What is of particular interest in Fig. 4.22 is that, for both ^{192,194}Pt nuclei, the relative location of the quasi-β-band head 0_2^+ energy is reproduced fairly well lying close to the 4_2^+ level. In addition, for ^{194}Pt, the present calculation suggests that the $0_2^+ \to 2_2^+$ E2 transition is dominant over the $0_2^+ \to 2_1^+$ E2 transition, which, although there is quantitative deviation, agrees with the experimental trend. The reason why such a quantitative difference occurs may be discussed in the future. Compared to the experimental data, the theoretical quasi-γ band is rather stretched and the band head 2_2^+ energy is somewhat large. The calculated 0_2^+ energy is also higher than the experimental one in particular for ^{192}Pt. Accordingly, the theoretical $B(E2; 2_2^+ \to 2_1^+)$ value is much smaller than experimental value with respect to the $B(E2; 2_1^+ \to 0_1^+)$ value. The deviations occur due to the derived κ value, which is somewhat larger than the phenomenological one [58]. For relatively high-lying side-band 2_3^+ and 4_3^+ energies, the calculated results may not seem to be much reliable, because even the ordering of these levels are not reproduced for ^{192}Pt.

Brief Summary

To summarize Sect. 4.3.1, spectroscopic calculations have been carried out, for the Pt isotopic chain in terms of the Interacting Boson Model Hamiltonian derived microscopically based on the (constrained) Hartree-Fock-Bogoliubov approach with the Gogny-D1S Energy Density Functional.

The Gogny-HFB calculations provide the total energy surface, which reflects, to a good extent, many-nucleon dynamics of surface deformation with quadrupole degrees of freedom and structural evolution in a given isotopic chain. By following the procedure proposed in Ref. [53], the energy surface of the Gogny-D1S EDF is mapped onto the corresponding bosonic energy surface, and can be then utilized as a guideline for determining the parameters of the IBM Hamiltonian. This enables one to calculate the spectroscopic observables with good quantum numbers (i.e., the angular momentum and the particle number) in the laboratory system without adjustment of levels.

4.3 Prolate-Oblate Shape Dynamics

By this approach, global tendencies of the experimental low-lying spectra of $^{172-200}$Pt nuclei are reproduced quite well not only for ground-state but also for side bands of mainly open-shell nuclei. It has been shown that shape/phase transition occurs quite smoothly from prolate to oblate deformations as a function of N in the considered nuclei $^{186-192}$Pt, where the γ instability plays an essential role. From the analysis in Fig. 4.18, the change of the mapped IBM energy surfaces in γ direction has been more vividly seen than in β direction, similarly to the corresponding Gogny-HFB energy surfaces. This is consistent with the conclusions in our earlier work [50] and also with many others along the same line. We have shown that the calculated spectra and the $B(E2)$ ratios behave consistently with the evolution of the topologies of the mapped energy surfaces and with the systematics of the derived IBM parameters as functions of the neutron number N. These derived parameters are qualitatively quite similar to the existing phenomenological IBM studies [56, 58]. By studying the level schemes in detail in comparison with the available experimental data, the present calculation agrees with the data fairly nicely and reflects the algebraic aspects of the IBM, e.g., the $\Delta \tau = \pm 1$ selection rule of the E2 decay patterns. We have also made predictions on some E2 transition patterns. These behaviors of the $B(E2)$ may need to be examined experimentally particularly for lighter, $A \leq 190$ nuclei with which there is currently few available data.

The evolution of ground-state shape as a function of both N and Z has been studied within neighboring isotopic chains such as Os, W, Hf and Yb [46]. It should be then of interest to study how the corresponding spectra and transition probabilities behave. More systematic spectroscopic analyses of these nuclei will be presented in Sect. 4.3.3, where the present mapping procedure will be applied more extensively using the new Gogny parametrization D1M.

On the other hand, the IBM Hamiltonian of Eq. (2.41) has rather simple form consisting of single-d-boson operator and the quadrupole-quadrupole interaction between proton and neutron bosons. The results provided by the present Hamiltonian were shown to be already quite promising. However, more studies may be necessary for further refinement, e.g., in describing detailed structure of the quasi-γ band, and for the description of configuration mixing. To these ends inclusions of cubic terms and intruder configuration will be also of great interest, whereas regarding the former the issue of triaxiality in the IBM will be studied in Chap. 6.

4.3.2 Evidence for Critical Points

The neutron-rich W, Os and Pt nuclei with $A \sim 190$–200 exhibit a very challenging structural evolution, which has already been extensively studied [97–104]. As originally pointed out in [98] the ratio $E_{4_1^+}/E_{2_1^+}$ in ^{190}W is anomalously small compared with the one in neighbouring isotopes. The most recent experimental data on the neutron-rich tungsten chain from ^{188,190,192}W [97, 100, 102] all suggest a change from a well deformed, axially symmetric prolate shape for lighter tungsten isotopes, to a more gamma-soft system for ^{190}W. This transition from a prolate to very gamma-soft system for neutron number $N = 116$ (i.e., for ^{190}W) is consistent with the recent observation of the second 2^+ state in ^{190}W which appears to lie lower than the yrast 4^+ in this nucleus [100]. The neutron-rich nature of the heavier W and Os nuclei make them experimentally challenging to study. However, in recent years, there has been some progress in their structural investigation following multi-nucleon transfer [97, 102, 103] and isomer and/or beta-delayed gamma-ray spectroscopy following projectile fragmentation reactions [98, 100, 101]. The current experimental information is limited to the yrast sequence in ^{190}W [100, 102] and the identification of the 2_1^+ state in ^{192}W [100]. It is interesting to note that the yrast 2^+ states in the $N = 116$ isotones ^{190}W and ^{192}Os have almost identical energies ($\sim$206 keV), as do the $N = 118$ isotones ^{192}W and ^{194}Os ($\sim$218 keV).

From the theoretical side, mean-field calculations have been performed which predict the shapes of these systems both with (e.g., [105]) and without (see, e.g., Refs. [46, 50], and references therein) the assumption of axial symmetry in the nuclear mean field. The IBM has also been applied to fit the spectral properties of W isotopes in a phenomenological way [57]. More recently and as described in Sect. 4.3.1 in this thesis, spectroscopic calculations have been carried out [72] to describe the structural evolution in Pt isotopes with the Gogny-D1S EDF [47]. In this section, we shall review the current spectroscopy relevant to the prolate-to-oblate shape/phase transition in neutron-rich Os and W isotopes. We also report, for the first time, the predicted excitation spectra and the transition probabilities on the neutron-rich Os and W nuclei. The spectroscopic calculations have been carried out in terms of the IBM Hamiltonian derived by mapping (constrained) Hartree-Fock-Bogoliubov (HFB) calculations, based on the Gogny-D1S EDF, using a similar technique as in [72] or in Sect. 4.3.1.

Mapped IBM Energy Surface and the Derived Parameters

Figure 4.23 shows the mapped IBM energy surfaces for $^{190-196}$Os and $^{188-194}$W nuclei up to 2 MeV excitation from the energy minimum. The corresponding HFB energy surfaces have been reported in Fig. 3 of Ref. [46]. The energy surfaces for both Os and W nuclei show similar tendencies. There are quantitative differences between the Pt and Os-W isotopic chains, namely that the topology of the energy surface changes more slowly in the former [72], compared to the latter in Fig. 4.23. An (almost) axially symmetric, oblate minima is observed in Pt nuclei with $N = 114$–120 and shallow triaxiality for $N = 110$ and 112 [46, 72]. On the other hand, the Os and W isotopes are predicted to have the corresponding oblate minima only for $N = 118$ and 120, with a more rapid change to axially symmetric prolate deformation for $N \leq 114$. Indeed, shallow triaxiality (i.e., γ-softness) appears only around $N = 116$ for both Os and W nuclei [46]. The corresponding mapped IBM energy surfaces reproduce these trends of the HFB energy surfaces of [46] well, whereas the location of the minimum in the IBM energy surface differs from that of the HFB energy surface of Ref. [46] in some nuclei as the presently used

4.3 Prolate-Oblate Shape Dynamics

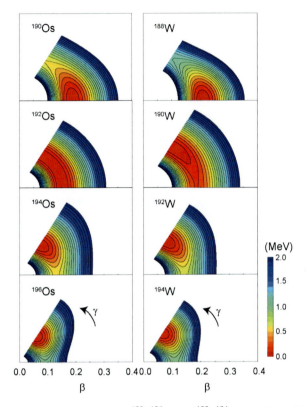

Fig. 4.23 Mapped IBM energy surfaces for $^{190-196}$Os and $^{188-194}$W nuclei up to 2 MeV excitation from the energy minimum within the ranges $0° \leq \gamma \leq 60°$. The energy surfaces are shown in terms of the fermionic deformation parameters β ($= \beta_B/C_\beta$) and γ ($= \gamma_B$). β_B and γ_B stand for the deformation parameters in the boson system and the numerical coefficient C_β was defined in Eq. (2.47). The figure is taken from Ref. [106]

IBM Hamiltonian, the one in Eq. (2.46), does not produce a triaxial minimum on the energy surface. The mapped energy surface for the $N = 116$ isotone, ^{192}Os is predicted to be very flat along the γ-direction. Similarly, the IBM energy surface for ^{190}W is also very flat, with the global energy minimum corresponding to a quadrupole deformation of $\beta \sim 0.15$ on the oblate side. This flatness is the consequence of the χ_π and χ_ν parameter values, such that their sum is close to zero. Comparing Os and W isotopes with the same neutron number, the W nuclei are generally steeper in both β and γ directions than the corresponding Os isotone. A similar trend is also observed in the corresponding HFB energy surfaces [46].

Figure 4.24 shows the evolution of the derived IBM parameters for the considered Os and W nuclei as functions of the neutron number N. The parameter values for Pt nuclei, taken from Ref. [72], are also shown for comparison. There are significant differences in quantitative details of the derived IBM parameter values between Os-W and Pt nuclei. In particular, the values of the parameter ε in Fig. 4.24a for Os and W

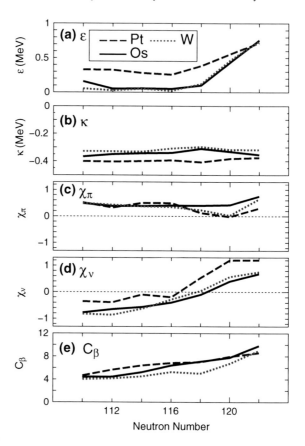

Fig. 4.24 Derived IBM parameter values for the considered Os and W nuclei, represented by *solid* and *dotted curves*, respectively, as functions of N. Results for Pt isotopes taken from Ref. [72] are also depicted for comparison. The figure is taken from Ref. [106]

nuclei are rather small in the region away from the shell closure as compared to Pt nuclei. In Fig. 4.24b, the magnitude of the parameter κ is larger than the analogous results for the Pt isotopes. The behavior of the parameters ε and κ is reflective of the HFB energy surfaces for Os and W nuclei being somewhat steeper in the β degree of freedom compared to the Pt isotopes, as discussed in Ref. [46]. The $\chi_{\pi,\nu}$ parameters in Fig. 4.24c, d (as well as their sum) behave similarly to those of Pt nuclei. For both $N = 110, 112$ the sum is almost zero in Pt isotopes while it is small for Os and W ones, but has a negative sign. This indicates a weak prolate deformation in the latter as seen in Fig. 4.23. In other words, the γ-soft structure is rather sustained in these Pt isotopes, but it is not for the corresponding Os and W isotopes. As in Fig. 4.24e, the scale parameter C_β in the present case behaves similarly as for the Pt nuclei with about the same order of magnitude.

4.3 Prolate-Oblate Shape Dynamics

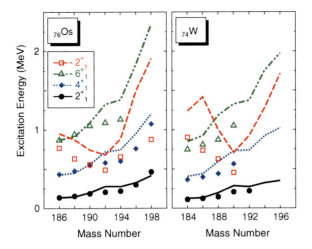

Fig. 4.25 Low-lying g.s. band and quasi-γ-bandhead (2_γ^+) energies (*curves*) for $^{186-198}$Os and $^{184-196}$W nuclei. Experimental data (*symbols*) are taken from Refs. [12, 59, 100]. The Gogny-D1S EDF is used. The figure is taken from Ref. [106]

Calculation of the Energy Spectra and $B(E2)$ Values

Using the derived parameters, we calculate excitation spectra and reduced E2 transition probabilities $B(E2)$. To do this, the Hamiltonian of the type defined in Eq. (2.41) is diagonalized as it has been done.

Figure 4.25 shows ground-state (g.s.) band and the quasi-γ-bandhead 2_2^+ (denoted by 2_γ^+) energies for Os and W isotopes. In general, the calculated results follow the experimental trends reasonably well, particularly for 2_1^+ energy. What is of interest in Fig. 4.25 is the behavior of the 2_γ^+ energy, exhibiting a kink for both ^{192}Os and ^{190}W. The experimental 2_γ^+ energy in ^{192}Os is lower than the 4_1^+ one. This is an evidence that the $A = 192$ nucleus is the most γ-unstable one among other Os isotopes. The present calculation follows the trend for Os isotopes well, and predicts a similar one for W isotopes exhibiting, however, more rapid change as a function of N. The location of the 2_γ^+ state for ^{196}Os (^{192}W) has not yet been fixed experimentally but the present calculations suggest that the 4_1^+ state is lower than the 2_γ^+ one in both ^{196}Os and ^{192}W. The calculated 2_γ^+ energy is generally higher than the experimental one, whereas the qualitative feature of experimental level is reproduced well.

Now we turn to the analysis of $B(E2)$ systematics, relevant to the considered low-lying states. The $B(E2)$ strength has been defined in Eq. (2.49). For the E2 operator of Eq. (2.50), we assume $e_\pi = e_\nu$, for simplicity, and discuss ratios of $B(E2)$s rather than their absolute values and the quadrupole moments for the corresponding excited states. Note that the $B(E2)$ ratio at each dynamical symmetry limit, shown below, means the one with infinite boson number [94].

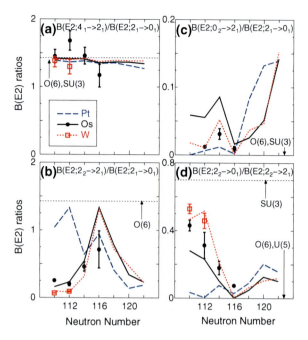

Fig. 4.26 Theoretical (*curves*) and experimental (*symbols* with error bar) [76–91] $B(E2)$ ratios for Os and W isotopes as functions of N. Theoretical results for Pt isotopes taken from Ref. [72] are also depicted as *dashed curves*, for comparison. Gogny-D1S EDF is used. The figure is taken from Ref. [106]

From Fig. 4.26a, we observe that the ratio $R_1 \equiv B(E2; 4_1^+ \to 2_1^+)/B(E2; 2_1^+ \to 0_1^+)$ does not change much, being close to its O(6) limit of IBM 10/7 (which is also the SU(3) limit of R_1). This trend persists for $N \geq 118$ where there is currently no available data. The ratio $R_2 \equiv B(E2; 2_2^+ \to 2_1^+)/B(E2; 2_1^+ \to 0_1^+)$, shown in Fig. 4.26b, is of particular interest as one can observe a significant difference in its value for the Pt and Os-W isotopes. The magnitude of the R_2 ratio is arguably the most appropriate and sensitive fingerprint for γ softness [72]. The R_2 values for both Pt and Os-W are relatively large and close to the O(6) limit (=10/7) for $N = 114$–118, where the nuclei show notable γ instability. For Pt nuclei, this trend persists even for $N \leq 112$, while smaller values are suggested for Os and W nuclei. These differences between the Pt and Os-W chains reflect the difference in the topology of the energy surface. The results for Os nuclei follow the experimental trend, which increases for $N = 110$–116. The present calculation for Os nuclei suggests the decrease of the R_2 value for $N \geq 118$, which corresponds to a suppression of γ softness. The ratio $R_3 \equiv B(E2; 0_2^+ \to 2_1^+)/B(E2; 2_1^+ \to 0_1^+)$ in Fig. 4.26c generally has a predicted value which is rather small, being close to zero (corresponding to the O(6) and SU(3) limits), as compared to R_1 and R_2 values. Note that the scale of the vertical axis in Fig. 4.26c is different from that of Fig. 4.26a, b. No rapid change with N is seen for R_3 as in R_2. Nevertheless, we should note the quantitative differences between the Pt and the Os-W nuclei. The branching ratio $R_4 \equiv B(E2; 2_2^+ \to 0_1^+)/B(E2; 2_2^+ \to 2_1^+)$

4.3 Prolate-Oblate Shape Dynamics

in Fig. 4.26d for Os follows the experimental trend for $N = 110$–116. The decrease of R_4 value from $N = 110$, close to 7/10 (SU(3) limit), toward $N = 116$, close to zero (O(6) and U(5) limit), reflects the corresponding structural evolution. The R_4 value for the Pt chain is close to zero, while for the Os-W chains, there is a significant change at $N = 116$. For the W nuclei, the ratio R_4 increases more rapidly than for the Os chain from $N = 116$ to 112. Earlier phenomenological studies suggested a similar increase [57].

Finally, we present in Fig. 4.27 the level schemes corresponding to the neutron-rich nuclei ^{190,192,194,196}Os and ^{190,192,194}W taken as representative samples. For ^{190,192}Os, for which there are significant experimental data, not only the g.s. band but also both the quasi-γ-bandhead 2_γ^+ and the quasi-β-bandhead 0_2^+ (denoted by 0_β^+) energies are reproduced quite well by the current calculations, although the detailed 'in-band' energy staggering looks different between the calculated and the experimental levels. The calculated $B(E2)$ values for ^{190,192}Os have been normalized to the experimental [76–91] $B(E2; 2_1^+ \to 0_1^+)$ value. Some algebraic feature is also apparent in the calculated results. The $\Delta\tau = \pm 1$ rule for the E2 decay pattern at the O(6) limit [96], (i.e., the dominance of $2_2^+ \to 2_1^+$ ($0_2^+ \to 2_2^+$) over $2_2^+ \to 0_1^+$ ($0_2^+ \to 2_1^+$)) in the present calculation also compares well with the experimental decay pattern.

The experimental value of the 2_1^+ energy for ^{192}Os is very close to that of its isotone ^{190}W (i.e., $E \approx 207$ keV). Also, the excitations energies of the 2_1^+ levels in these isotones are also quite similar to each other. The present calculations reproduce this overall trend well. In fact, the calculated $E(2_1^+) = 0.280$ (0.278) MeV and 0.286 (0.274) MeV for ^{192}Os (^{194}Os) and ^{190}W (^{192}W) nuclei, respectively. For ^{192}Os and ^{190}W nuclei, the calculated g.s. band energies are rather stretched, and the 2_γ^+ energies are in good agreement with the respective experimental data. In the calculated quasi-γ band of ^{190}Os and ^{192}Os nuclei, one observes a staggering as 2_γ^+ (3_γ^+ 4_γ^+) (5_γ^+ 6_γ^+), ... etc. By contrast, the experimental energy spacing shows a more regular pattern. This deviation may be related to the topology of the mapped IBM energy surface in Fig. 4.23, which is flat in γ direction, while the corresponding Gogny-D1S energy surface exhibits shallow triaxial minimum [46]. Some additional interaction term, such as the cubic term [15], may need to be introduced in the boson Hamiltonian to correct the deviation for detailed structure of quasi-γ band. This issue will be addressed in Chap. 6.

For ^{194,196}Os nuclei, the predicted 2_1^+ and 4_1^+ energies reproduce the experimental ones. The quasi-β-bandhead energy for ^{194}Os in the present calculation is notably larger than the experimental value, which is a consequence of the peculiar topology of the Gogny-HFB energy surface, which exhibit a pronounced oblate minimum with a relatively small deformation. This results in the larger value of the parameter κ than the one in the IBM phenomenology [57] which would give good agreement for the excited 0^+ energies. The positions of the 2_γ^+ and the 0_β^+ energies for ^{196}Os are predicted to lie below and beyond the 6_1^+ level, respectively. For the exotic ^{192}W and ^{194}W nuclei, the present calculation suggests a quite similar level pattern to their respective isotones, ^{194}Os and ^{196}Os.

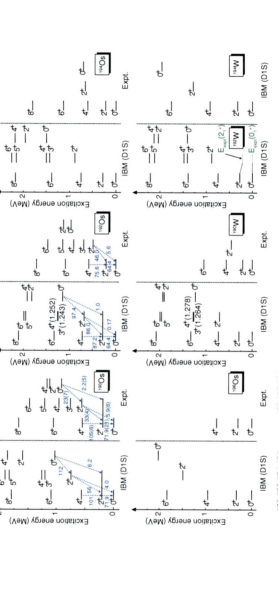

Fig. 4.27 Level schemes for ^{190,192,194,196}Os and ^{190,192,194}W nuclei. The theoretical B(E2) values (in Weisskopf units) for ^{190,192}Os are normalized to the experimental [76–91] B(E2;$2_1^+ \to 0_1^+$) value. Calculated 3_1^+ and 4_2^+ energies for ^{192}Os and ^{190}W are shown (in MeV units) in the parentheses because otherwise these energies look identical. Note that, for ^{192}W, experimental data are shown in the same panel as theoretical ones. Gogny-D1S EDF is used. The figure is taken from Ref. [106]

4.3.3 Systematics from Gogny-D1M Functional

Next we extend the analyses done in Sects. 4.3.1 and 4.3.2 over very neutron-rich Yb, Hf, W, Os and Pt isotopes, for some of which the experimental data are quite scarce but are expected to be available. An additional motivation is to explore some possibilities to refine the predictive power of the method for the considered mass region.

Starting from the constraint HFB theory with Gogny-D1S [47] EDF, the method of [53] was applied to the spectroscopy of Pt isotopes [72], and some of Os and W isotopes [106]. In this paper we present more systematic analyses, extending over the neighboring Hf and Yb nuclei. While the D1S parametrization has been considered as global and able to describe many low-energy nuclear data with reasonable predictive power (see, for example, Refs. [46, 50, 54] and references therein), we here take the Gogny-D1M functional [52] throughout. The first systematic explorations [50, 52], including odd nuclei within the equal-filling approximation [107–109], suggest that the new incarnation D1M of the Gogny-EDF essentially keeps the same nuclear structure predictive power as the (standard) parametrization D1S, but still further work should be in order. At a number of places in this section, the results are compared with those with the standard D1S functional, already presented in Sects. 4.3.1 and 4.3.2.

Since well deformed axially symmetric nuclei should be treated in this section, the IBM-2 Hamiltonian of Eq. (3.3), which contains the LL term, is employed:

$$\hat{H}_{\text{IBM}} = \varepsilon \hat{n}_d + \kappa \hat{Q}_\pi \cdot \hat{Q}_\nu + \alpha \hat{L} \cdot \hat{L}, \tag{4.7}$$

where definition of each term in the above Hamiltonian was already shown in Chap. 3.

The parameters involved in the first two terms of the Hamiltonian $\hat{H}_{\text{IBM}}$ in Eq. (4.7), ε, κ, $\chi_{\pi,\nu}$ and C_β, are fixed using the fitting method of Ref. [11]: the Gogny-HFB energy surface $E_{\text{HFB}}(\beta, \gamma)$ is mapped onto the corresponding point of the IBM-2 energy surface $E_{\text{IBM}}(\beta_B, \gamma_B)$ so that the latter becomes identical to the former as much as possible.

However, the LL term contributes to the energy surface in the same way as the d-boson number operator, but with a different coefficient, 6α. Hence one cannot fix the α value only by the mapping of the energy surface. One should then go beyond the basic energy-surface analysis, which is considered as a zero-frequency mode, to take into account the rotational cranking, which on the other hand takes on a specific non-zero frequency feature. The α value is determined with the procedure of Ref. [110], where the cranking moment of inertia was compared between nucleon and boson systems.

We then calculate the moment of inertia for the 2_1^+ excited state by the Thouless-Valatin (TV) formula [111],

$$\mathscr{I}_{\text{TV}} = 3/E_\gamma. \tag{4.8}$$

Here, E_γ stands for the 2_1^+ excitation energy obtained from the self-consistent cranking method with the constraint $\langle \hat{J}_x \rangle = \sqrt{L(L+1)}$, where $\hat{J}_x$ represents the

x-component of the (fermion) angular momentum operator [46]. In the original work of Ref. [110], we used the Inglis-Belyaev formula [112, 113], which is valid for the rotational regime, but the present TV moment of inertia is more general.

For the boson system, we calculate the moment of inertia for the intrinsic state, denoted by $\mathscr{J}_{\text{IBM}}$, using the cranking formula of Ref. [114] as a function of the deformation parameters β_B and γ_B. $\mathscr{J}_{\text{IBM}}$ is written as

$$\mathscr{J}_{\text{IBM}}(\beta_B, \gamma_B) = \lim_{\omega \to 0} \frac{1}{\omega} \frac{\langle \Phi(\beta_B, \gamma_B) | L_x | \Phi(\beta_B, \gamma_B) \rangle}{\langle \Phi(\beta_B, \gamma_B) | \Phi(\beta_B, \gamma_B) \rangle}, \tag{4.9}$$

where ω and L_x stand for the cranking frequency and the x-component of the boson angular momentum operator, respectively.

$\mathscr{J}_{\text{IBM}}$ involves six parameters ε, κ, χ_π, χ_ν, C_β and α. All these parameters but α are already fixed by the energy-surface analysis. The α value for each nucleus is obtained so that the $\mathscr{J}_{\text{IBM}}$ value at the equilibrium point where the boson energy surface $E_{\text{IBM}}(\beta_B, \gamma_B)$ is minimal becomes identical to the $\mathscr{J}_{\text{TV}}$ value at its corresponding energy minimum.

The values of all derived IBM-2 parameters are summarized in Table 4.1. When diagonalizing the Hamiltonian Eq. (4.7), the ε parameter is shifted by $\Delta \varepsilon = 6\alpha$. The ε value listed in Table 4.1 is the one with this shift. With the parameters listed in Table 4.1, the diagonalization of the Hamiltonian in Eq. (4.7) the collective spectra and the E2 transition rates.

Energy Surfaces

In comparison to the original microscopic energy surfaces in Fig. 4.28 for Yb-Pt isotopes with $112 \leq N \leq 120$ obtained from the Gogny-D1M functional, Fig. 4.29 shows the mapped IBM-2 energy surfaces. Each energy surface is plotted up to 2 MeV from its absolute minimum. Note that the original Gogny-D1M energy surfaces are not shown as they are rather similar in topology to those resulting from the D1S functional reported previously [46].

For all the isotopes but the Pt, the energy minimum shifts from the prolate ($\gamma = 0°$) to the oblate ($\gamma = 60°$) sides as the number of neutrons increase, passing through the most notable γ-soft nuclei $N = 116$. The derived χ_π and χ_ν values for many $N = 116$ nuclei then satisfy $\chi_\pi + \chi_\nu \approx 0$, as summarized in Table 4.1. This choice of the χ parameters in the IBM-2 is an origin of the almost totally flat topology of the energy surface, as seen for example in ^{192}Os in Fig. 4.29. The change in the topology of the energy surface is an evidence of prolate-to-oblate shape/phase transition, which becomes sharper for smaller Z. The Gogny-D1S energy surfaces reported in [46, 72] were somewhat steeper in both β and γ directions than the present Gogny-D1M calculation.

A difference is apparent between the energy surfaces of the Pt isotopes and those of others. For the Pt, the variation of the energy surface takes place much moderately. Such slow structural transition in Pt isotopes was also observed in the case of the

4.3 Prolate-Oblate Shape Dynamics

Table 4.1 The derived parameters of the IBM-2 Hamiltonian $\hat{H}_{\text{IBM}}$ of Eq. (4.7)

Nuclei	ε (keV)	$-\kappa$ (keV)	$\chi_\pi \times 10^3$	$\chi_\nu \times 10^3$	α (keV)	C_β
^{180}Yb	212	265	337	−991	−9.06	3.60
^{182}Yb	169	265	300	−900	−11.4	3.70
^{184}Yb	279	271	302	−548	−9.84	3.87
^{186}Yb	418	268	147	−106	−9.54	4.90
^{188}Yb	528	265	418	43	−4.68	5.13
^{190}Yb	769	267	332	573	−0.185	5.50
^{192}Yb	806	271	461	862	21.5	7.20
^{182}Hf	124	280	489	−913	−5.61	3.93
^{184}Hf	128	282	458	−938	−8.01	4.07
^{186}Hf	109	275	400	−700	−4.85	4.40
^{188}Hf	250	277	282	−208	−7.90	5.30
^{190}Hf	442	280	403	−30	−5.99	5.48
^{192}Hf	619	273	388	443	2.79	5.94
^{194}Hf	716	277	534	805	18.4	8.20
^{184}W	50.4	286	409	−859	−0.400	4.09
^{186}W	36.8	285	389	−835	−2.30	4.50
^{188}W	69.6	289	401	−662	−1.44	4.80
^{190}W	71.3	275	572	−419	−2.72	5.60
^{192}W	231	270	189	147	−4.15	6.30
^{194}W	627	291	392	536	−5.74	6.87
^{196}W	686	281	745	822	15.3	8.50
^{186}Os	142	310	331	−689	−0.433	4.40
^{188}Os	162	318	352	−672	−2.78	4.83
^{190}Os	86.7	303	412	−509	−2.61	5.40
^{192}Os	91.5	292	502	−488	−3.09	6.15
^{194}Os	289	305	401	−77	−6.04	6.74
^{196}Os	541	298	336	513	−5.94	7.64
^{198}Os	683	304	573	793	8.50	9.66
^{188}Pt	187	328	409	−487	8.16	4.81
^{190}Pt	215	336	300	−10	5.93	5.56
^{192}Pt	311	362	265	44	−0.117	6.44
^{194}Pt	312	366	490	−50	0.214	6.85
^{196}Pt	435	356	475	311	1.87	7.28
^{198}Pt	489	319	611	565	8.80	7.90
^{200}Pt	719	308	467	949	−4.69	8.78

These values are taken from Ref. [115]

D1S functional [50, 72]. While a certain quantitative difference occurs between the two Gogny functionals the conclusion does not change.

It should be noted that the Gogny-HFB calculation suggests shallow triaxiality for the transitional, $N = 116$ Os and W nuclei [46]. In contrast, the mapped IBM-2 energy surfaces in Fig. 4.29 are flat in γ, as the only γ-dependent term of the bosonic surface is proportional to $\cos 3\gamma$. This is the case as long as the boson Hamiltonian is

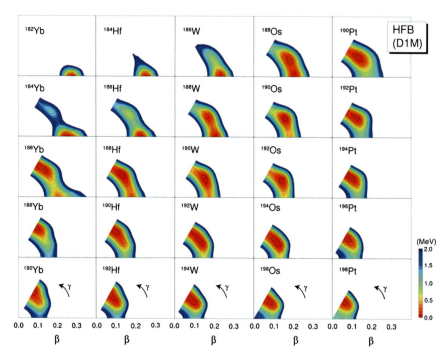

Fig. 4.28 The Gogny-D1M energy surfaces, depicted within $0 \le \beta \le 0.4$ and $0° \le \gamma \le 60°$ up to 2 MeV excitation from the minimum. Contour spacing is 100 keV

composed of the one and the two-body interactions. It is only a three-body (so-called cubic) term that creates rather stable minimum different from $\gamma = 0$ and $60°$. This issue will be revisited.

Correlation Energies

We next discuss the signature of shape transition from a simple perspective. To do this we consider the following quantity that will be called *correlation energy* hereafter,[2] which was introduced also in Ref. [11]:

$$E_{\text{Corr}} = E_{\text{IBM}}(0_1^+) - \langle \hat{H}_{\text{IBM}} \rangle_{\text{min}}, \qquad (4.10)$$

where the first term $E_{\text{IBM}}(0_1^+)$ is the eigenenergy of the IBM-2 Hamiltonian, Eq. (4.7), for the $L^\pi = 0^+$ ground state, and the second term $\langle \hat{H}_{\text{IBM}} \rangle_{\text{min}}$ denotes the minimum

---

[2] The overall systematic trend of the correlation energy and its effect on the ground-state properties will be revisited in Chap. 7 with more ample definitions and references. However, the discussion about the correlation energy in this section does not require prior reading of Chap. 7.

4.3 Prolate-Oblate Shape Dynamics

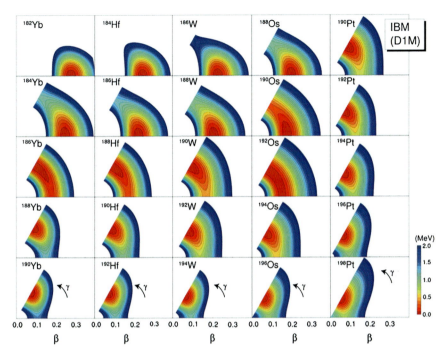

Fig. 4.29 Same as Fig. 4.28, but for the IBM-2 energy surfaces, obtained by the mapping from the Gogny-D1M energy surface. Figure is taken from Ref. [115]

value of the IBM-2 energy surface, that is obtained by the variation with respect to β and γ.

In the self-consistent mean-field calculation with a given EDF (e.g., Ref. [116]), the quantum-mechanical effect similar to the one defined in Eq. (7.4) can be also extracted by comparing the minimum value of the total energy surface of the mean field with the $L^\pi = 0^+$ eigenenergy resulting from the restoration of the broken symmetries and the configuration mixing.

In the present study, the boson Hamiltonian is diagonalized in a set of bases which have the angular momentum and the particle number as good quantum numbers. Thus the quantity defined in Eq. (7.4) appears to involve the effects relevant to the angular momentum projection on the state with good L plus the configuration mixing, and can be thus equivalent to the correlation energy normally considered in the studies like the GCM.

E_{Corr} depends rather significantly on the underlying shape transition. Figure 4.30 shows that for each considered isotopic chain the correlation energy is maximal in its magnitude at the neutron number $N \sim 116$, which corresponds to the transition point of the prolate-oblate transition, and that decreases as the neutron shell closure $N = 126$ is approached. This is correlated strongly with the systematics of the energy surface in Fig. 4.29. For Pt isotopes, the magnitude of E_{Corr} decreases with N,

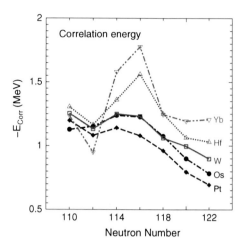

Fig. 4.30 Correlation energy in the ground state E_{Corr}, defined in Eq. (7.4), for the considered Yb, Hf, W, Os and Pt isotopes. The figure is taken from Ref. [115]

indicating that a clear transition is not expected for the nuclei with $110 \leq N \leq 122$. These are well known results [116, 117], and also seem to be quite consistent with the conclusion made in our previous work of Ref. [11] for Sm isotopes.

When compared with the analysis by the GCM configuration mixing using a Skyrme functional [116] for the same mass region as considered here, the magnitude of the present correlation energy E_{Corr} is rather small, whereas the qualitative features mentioned above do not contradict the GCM results.

When the total energy needs to be treated, one has to explicitly include in the IBM-2 Hamiltonian Eq. (4.7) a global term that is constant for a nucleus. This global term can be fixed using the minimum value of the HFB energy surface and that of the IBM-2 energy surface. In comparison to some rare-earth nuclei such as Nd-Sm-Gd isotopes, where a distinct first-order shape transition is observed [2], the shape transition occurs rather moderately in the considered mass region. Thus, it can be shown that, contrary to E_{Corr} in Fig. 4.30, any drastic change with nucleon number is not expected in some other quantities in the ground-state, like two-nucleon separation energies.

Moment of Inertia in the Cranking Formula

Based on the analysis in Sect. 4.3.3, we discuss the moment of inertia to see more about the correlation effect involved by the diagonalization of the boson Hamiltonian. The effect is most nicely illustrated in W isotopes, for which relatively many experimental spectroscopic data exist.

We show in Fig. 4.31 the moments of inertia of W isotopes, calculated by the cranking formula for the boson intrinsic state $\mathcal{J}_{\text{IBM}}$ in Eq. (4.9) and those taken from the 2_1^+ excitation energies of the IBM-2 and of the experiments [12] using the rotor formula $L(L + 1)$. Note that the cranking moment of inertia of IBM-2 is, due

4.3 Prolate-Oblate Shape Dynamics

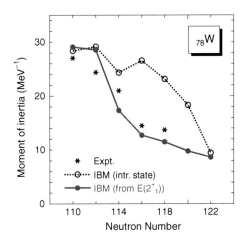

Fig. 4.31 Moments of inertia of W isotopes, computed by the cranking formula for the boson intrinsic state (intr. state), by the the rotor formula $L(L+1)$ using the 2_1^+ excitation energies of the IBM-2 and of the experimental data [12]. The figure is taken from Ref. [115]

to the correction by the LL term, set identical to the Thouless-Valatin (TV) moment of inertia. Thus the TV moment of inertia is not depicted in Fig. 4.31.

The experimental moment of inertia decreases with N and the slope of this decrease seems to change suddenly at $N = 116$. This change suggests a gradual shape transition, which is possibly of second-order type. The moment of inertia of the intrinsic state, in contrast, decreases smoothly with the exception of the kink at $N = 114$. Perhaps such a kink reflects a detailed shell structure irrelevant to the present work. However, once the correlation effect is taken into account by the diagonalization, the kink disappears and theoretical moment of inertia falls on the same systematics as the experimental data.

It appears that, from Fig. 4.31, the cranking moment of inertia still works for the nuclei $N = 110$ and 112, for which one cannot see any difference from the moment of inertia taken from the energy of the 2_1^+ eigenstate. In the transitional region of $114 \leq N \leq 118$, where according to Fig. 4.30 large amount of correlation energy should be involved, however, the moment of inertia of the intrinsic state is far from sufficient and configuration mixing by the diagonalization of Hamiltonian becomes significant to describe the experimental trend.

Excitation Spectra

Next we discuss in Figs. 4.32 and 4.33 the low-lying excitation spectra for the considered isotopic chains. In Fig. 4.32 the 2_1^+, 4_1^+, 6_1^+ and 8_1^+ excited states in the ground-state band for the considered Pt (a), Os (b), W (c), Hf (d) and Yb (e) isotopes are displayed, while other levels not belonging to the ground-state band, $0_2^+, 2_2^+, 3_1^+, 4_2^+$ and 5_1^+, are shown in Fig. 4.33 for the Pt (a), Os (b), W (c), Hf (d) and Yb (e) isotopes.

Experimentally [12, 100, 101], excitation energies of the ground-state band shown in Fig. 4.32, namely the $2_1^+, 4_1^+, 6_1^+$ and 8_1^+ yrast states, increase as the neutron shell

112   4 Weakly Deformed Systems with Triaxial Dynamics

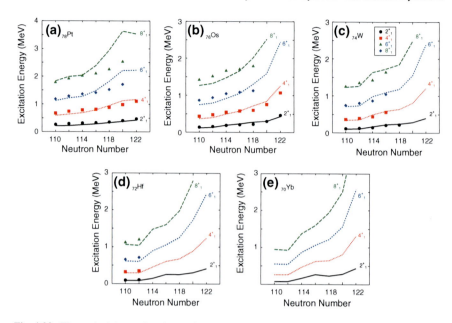

Fig. 4.32 Theoretical (*curves*) and experimental [12, 100, 101] (*symbols*) low-lying spectra of Yb, Hf, W, Os, and Pt isotopes with $110 \leq N \leq 122$ for the 2_1^+, 4_1^+, 6_1^+ and 8_1^+ states. Symbols for the experimental levels are defined in the panel **c**. The figures are taken from Ref. [115]

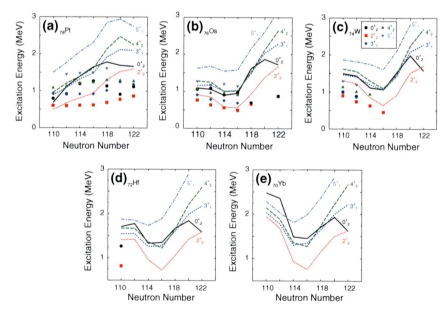

Fig. 4.33 Same as Fig. 4.32, but for the side-band states 0_2^+, 2_2^+, 3_1^+, 4_2^+ and 5_1^+. Each figure is taken from Ref. [115]

4.3 Prolate-Oblate Shape Dynamics

closure $N = 126$ is approached. These spectra become more compressed for the isotopic chains with smaller Z, when departing from the shell closure $Z = 82$. The present results follow the experimental isotopic trend for those nuclei with available data. The same systematics can be observed with the Gogny-D1S functional [72, 106].

When approaching the neutron shell closure $N = 126$, the present energy levels for Pt (Fig. 4.32a) and Os (Fig. 4.32b) nuclei overestimate the experiments. This is mainly because the derived κ values are too large in magnitude as compared to those used in phenomenological studies [57, 58]. This is a consequence of the weak oblate deformation for the corresponding Pt nuclei [46, 50]. In Refs. [72, 106], the D1S functional suggested the deviation of the yrast levels nearby the shell closure $N = 126$ similar to the present results for Pt, Os and W isotopes in Fig. 4.32a–c, respectively.

At the quantitative level, the LL term has a remarkable influence on the ground-state band. Without this term, the experimental yrast spectra would not be reproduced that precisely. This is particularly the case with lighter W (Hf) isotopes with $N = 110$ and 112, which follow the rotor formula $L(L+1)$ with their respective experimental ratios being $E_{4_1^+}/E_{2_1^+} = 3.27(3.29)$ and 3.23 (3.26) [12]. For these nuclei, the present results in Fig. 4.32c–d compare rather well with the experiments. We note that the analysis could be applied also for much lighter and more rotational Hf and Yb nuclei around the middle of the major shell.

We now turn to the description of the side-band energies. To begin with, the excited 0^+ (0_2^+) state is considered in Fig. 4.33. It is well known that the intruder configurations may play a role for Pt isotopes near the mid-shell where the oblate-prolate shape coexistence is observed [45, 63]. The phenomenological IBM study (see Ref. [67], for instance) considers particle-hole excitations across the $Z = 82$ proton shell. In this kind of work one needs to extend the boson model space as to take into account the intruder configuration with additional proton bosons, arising from (mainly) the $2p$-$2h$ excitation. The normal and the intruder configurations are mixed, and the model Hamiltonian should be then diagonalized in such enlarged configuration space. The validity of this mixing calculation has been discussed extensively [68, 69], and is thus of great interest.

The mixing in general becomes more significant around the middle of the major shell. In Fig. 4.33a, the present 0_2^+ excitation energies for $N \leq 116$ Pt isotopes, as well as those with Gogny-D1S [72], agree fairly well with the data, even without taking into account the mixing between normal and intruder states. This is the same conclusion as in Ref. [72], which took the Gogny-D1S functional. Furthermore, the original HFB energy surfaces for Pt isotopes do not exhibit clear coexisting minima. Due to this the present framework cannot fix the parameters for both the normal and the intruder configurations as well as those for the operators mixing the two configurations. Although such a mixing calculation is a rather subtle problem, it is very interesting to study the extent to which the intruder configuration plays a role when applied in the present mapping method.

It was shown experimentally [118–121] that, in the non-yrast states of lighter W, Os and Pt nuclei, the band mixing could arise more or less from the coexistence of different intrinsic states mentioned above, and makes it rather difficult to identify the clear band structure by a model prediction. The band-mixing feature should be outside of the model space of bosons with low-spin on which the IBM is built, and may be somewhat difficult to be reproduced. It is yet not clear whether the similar complicated band mixing is observed in the exotic Yb and Hf isotopes.

The 2^+_2 level, which is normally the bandhead of the $K^\pi = 2^+$ (so-called quasi-γ) band, is a good test for the evolving triaxiality in a given isotopic chain. Figure 4.33 shows that the calculated 2^+_2 level of $N = 116$ nucleus is lowest among each of Yb, Hf, W and Os isotopes. Experimental excitation energies keep steady (decrease) in Pt (Os, W) isotopes as N increases from 110 to 116.

The decrease of a set of energies for 2^+_2, 3^+_1, 4^+_2 and 5^+_1 states occurs more rapidly for lower Z isotopes, having larger number of active bosons. Around $N = 116$ a change in this tendency occurs and the excitation energies increase in the few cases measured. The present calculation follows this trend, particularly for the bandhead 2^+_2 energy in Os isotopes in Fig. 4.33b. Every γ-band energy turns to increase at $N = 116$ in all the considered isotopic chains but the Pt. This is nicely supported by the measurement in the Os isotopes, whereas the scale of 2^+_2 excitation energy in the calculation is generally overestimated for $N \geq 118$. The same holds for W isotopes. These tendencies, when looked at both in isotopic and isotonic chains, are also quite consistent with what can be expected from the systematics of the corresponding energy surfaces exhibiting the most prominent γ-softness at $N = 116$.

A remarkable difference between the theoretical and the experimental quasi γ-band structure observed in Pt and Os isotopes is that the 3^+_1 and the 4^+_2 states, and the 5^+_1 and the 6^+_2 states as well, in the calculations form doublets, which are absent in the data. As the states in Fig. 4.33a–b but the 0^+_2 one, are supposed to be the quasi-γ band states, formation of the doublets is an evidence of the γ-unstable [122] or O(6) dynamical symmetry [93, 94], in which the spectra belonging to the same family of the quantum number τ are nearly degenerated. Since the rigid triaxial rotor model with $\gamma = 30°$ [28] predicts the doublets $(2^+, 3^+)$, $(4^+, 5^+)$, etc., in the γ band, the experimental data in Fig. 4.33 for (a) Pt, (b) Os, and (c) W isotopes suggest a situation rather in between the γ-unstable rotor and the rigid-triaxial rotor pictures. The discrepancy of the γ-band energies occurs because the IBM-2 energy surface does not show the triaxial minimum which is, however, seen in the original HFB energy surface.

There are several possible effects which may eliminate this staggering in the γ-band spectra and improve the agreement with the experiments at the quantitative level. In the present paper, however, we do not look into the details of this issue due to the large number of additional parameters to be introduced and the lack of experimental data for the Yb and Hf nuclei. First, a three-body (cubic) term, which partially breaks O(6) symmetry, may correct the deviation. This has been done mainly in the IBM-1 [15, 16]. For the present case some type of cubic term appears to be necessary mainly for W, Os and Pt nuclei, where the Gogny HFB energy surface exhibits a shallow, but stable triaxial minimum [46]. While the calculated excitation

4.3 Prolate-Oblate Shape Dynamics

energies of the quasi-γ band for Yb and Hf in Fig. 4.33d, e look like that of pure O(6) limit as well, the validity of this term seems to be marginal in these cases. Indeed for the Yb and Hf isotopes the original Gogny-D1M energy surface indicates the discrete change of the minimum point from the oblate ($\gamma = 60°$) to the prolate ($\gamma = 0°$) sides, similarly to the Gogny-D1S energy surface [46].

The second possibility would be to relax the constraint on the deformation parameters γ_π and γ_ν so that they could take different values. As the IBM-2 can be viewed as a two-fluid system consisting of proton and neutron bosons, the phase-structure analysis would be exploited in the context of the coherent-state formalism [123, 124], whereas it is not obvious to define a consistent mapping procedure for realistic cases.

The third would be the inclusion of higher-spin bosons, like the g-boson. It is not independent of the first possibility involving the cubic term, since the cubic term can be derived effectively from the renormalization of the g boson into the sd-boson sector [15]. This would, of course, make the problem more complicated.

We now address why the side-band spectra, particularly for Pt in Fig. 4.33a and Os in Fig. 4.33b isotopes, are overestimated in the present calculation when approaching the $N = 126$ shell closure. The direct reason would be that mostly the microscopic Gogny energy-surface calculation predicts oblate deformation with small quadrupole moment but with rather large amount of deformation energy indicated by the depth of the potential minimum [46]. Such topology of the HFB energy surface is not well described by the IBM-2 Hamiltonian close to the end of the major shell $Z = 82$. Nearby the closed shell in general, one has relatively small number of bosons. The deviation of the spectra seems to be due to this limited degrees of freedom. The problem on the description of the side-band energies was observed in other cases of the shape transitions in the different mass regions [11, 53], and is still an open question. According to the above argument it may be expected that the predicted levels for exotic Yb and Hf isotopes in the vicinity of the shell closure $N = 126$ may be too high.

To examine the problem further, it is interesting to see the relevant energy ratios, as they nicely trace the shape transition. Figure 4.34 depicts the energy ratios (a) $R_{4/2} \equiv E(4_1^+)/E(2_1^+)$ and (b) $R_{4\gamma} \equiv E(4_1^+)/E(2_2^+)$ as functions of N. The ratio $R_{4/2}$ is the most basic and well-studied measure for the collective structural evolution. The ratio $R_{4\gamma}$ is considered in order to see the location of the bandhead of the quasi-γ band 2_γ^+ (2_2^+) relative to the 4_1^+ excitation energy, as this difference can help measure the degree of γ softness.

In Fig. 4.34a, the experimental $R_{4/2}$ ratio exhibits a monotonic decrease as a function of N from the rotor limit of $R_{4/2} = 3.3$ in the vicinity of $N = 110$ to the O(6) limit of $R_{4/2} = 2.5$. This obviously reflects the transition from the rotor to the γ-unstable shapes. Also of particular interest is the difference of the $R_{4/2}$ ratio between Pt isotopes and the other isotopes. The $R_{4/2}$ ratio for all the Pt isotopes studied remains practically constant all the way, being close to the O(6) limit of 2.5. The present calculation compares rather well with experimental $R_{4/2}$ values from $N = 110$ to 116 in Os and W isotopes, while an increase is suggested for $N \geq 118$, contrary to the experimental tendency of Os isotopes. This is the consequence of the

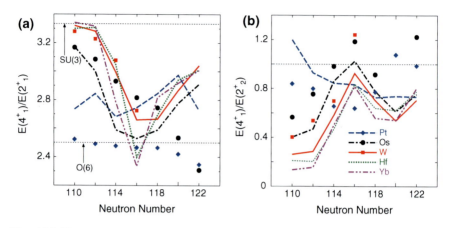

Fig. 4.34 Theoretical (*curves*) and experimental (*symbols*) [12] energy ratios (**a**) $R_{4/2} = E(4_1^+)/E(2_1^+)$ and (**b**) $R_{4\gamma} = E(4_1^+)/E(2_2^+)$ as functions of N. Definitions of the theoretical curves and the symbols for the experimental data appear in panel **b**. Both figures are taken from Ref. [115]

unexpectedly large χ_π and χ_ν values with positive sign, as seen in Table 4.1, since the corresponding IBM-2 energy surfaces exhibit notable oblate deformation.

The energy ratio $R_{4\gamma}$ appears in Fig. 4.34b. The experiment shows that in the lighter Pt, Os and W isotopes with $N = 110$, 112 and 114, the ratio is below unity. For Pt isotopes the ratio $R_{4\gamma}$ remains all the way with values close to unity, which is qualitatively reproduced by the present calculations. The experimental ratios for lighter Os-W isotopes increase with N and overpass $R_{4\gamma} = 1$ at $N = 116$. This behavior is also reproduced. From this point on there is a drastic change in the tendency and the ratios decrease from $N = 116$ to 118, being in good agreement with experiment. In the heavier isotopes with $N \geq 118$ there is a new tendency that the calculated ratio continues decreasing, being much below the unity, whereas the experimental ratio for Os isotopes keeps increasing, being larger than unity. For Yb and Hf, the data are not available, but similar tendencies to those for W and Os are predicted in the present calculation. The results presented here do not differ much from the case of D1S functional [72, 106].

B(E2) Systematics

In the last part of Sect. 4.3.3, we examine the $B(E2)$ systematics for a few essential cases of the shape transition. The $B(E2)$ ratios relevant to the bandhead of quasi-γ band, 2_2^+ state, can be stringent tests.

We show in Fig. 4.35 the $B(E2)$ ratio (a) $B(E2; 2_2^+ \to 2_1^+)/B(E2; 2_1^+ \to 0_1^+)$ and the branching ratio (b) $B(E2; 2_2^+ \to 0_1^+)/B(E2; 2_2^+ \to 2_1^+)$ for the considered isotopes in comparison with the data [76–91].

4.3 Prolate-Oblate Shape Dynamics

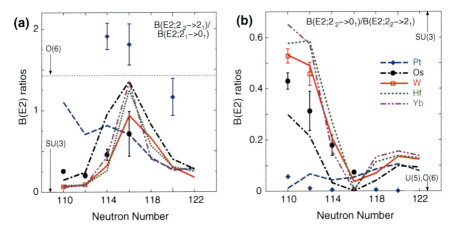

Fig. 4.35 The $B(E2)$ ratio (**a**) $B(E2; 2_2^+ \to 2_1^+)/B(E2; 2_1^+ \to 0_1^+)$ and the branching ratio (**b**) $B(E2; 2_2^+ \to 0_1^+)/B(E2; 2_2^+ \to 2_1^+)$ for relevant low-lying states of the considered Yb, Hf, W, Os, and Pt isotopes with Gogny-D1M EDF. Experimental data for W, Os, and Pt isotopes are taken from Ref. [76–91]. Definitions of symbols and theoretical curves appear in panel **b**. Both figures are taken from Ref. [115]

The $2_2^+ \to 2_1^+$ E2 transition is useful as a signature of the structural evolution involving the γ softness, since it in this case shows an enhanced sensitivity to the neutron number N. The $B(E2; 2_2^+ \to 2_1^+)/B(E2; 2_1^+ \to 0_1^+)$ ratios for Pt isotopes differ notably from those of other isotopes. For Yb, Hf, W, and Os isotopes, the calculated ratio is peaked at $N = 116$. This confirms that the $N = 116$ nuclei are soft in the γ direction for these isotopes. Particularly the relevant Os nucleus with $N = 116$ is closest to the O(6) limit of 10/7. On the other hand, for Pt isotopes the calculated $B(E2; 2_2^+ \to 2_1^+)/B(E2; 2_1^+ \to 0_1^+)$ value keeps increasing toward $N = 110$ to be close to the O(6) limit, rather than taking a maximum at $N = 116$. This tendency appears to be consistent with what are expected from the topology of the HFB energy surface and from the predicted systematics of the quasi-γ bandhead in Fig. 4.33a, which reflects that the γ softness persists for rather wide region in the Pt isotopic chain.

When compared with the D1S force [106], the present D1M result suggests that the ratio $B(E2; 2_2^+ \to 2_1^+)/B(E2; 2_1^+ \to 0_1^+)$ is rather sensitive to the isotopic chains. In fact, in Fig. 4.35a, the $B(E2; 2_2^+ \to 2_1^+)/B(E2; 2_1^+ \to 0_1^+)$ value at $N = 116$ appears to have a certain Z dependence when the D1M functional is used. For instance, the $B(E2; 2_2^+ \to 2_1^+)/B(E2; 2_1^+ \to 0_1^+)$ value for Yb isotopes is generally far from the O(6) limit for $110 \leq N \leq 120$, being much closer to the SU(3) limit of zero than Os isotopes. It has been noticed in Ref. [106], however, that the calculated value of this $B(E2)$ ratio is practically the same for Yb, Hf, Os, W isotopes when the D1S functional is taken. It should be interesting to test whether or not the notable Z dependence, as seen in the present D1M case, is observed experimentally.

The branching ratio $B(E2; 2_2^+ \to 0_1^+)/B(E2; 2_2^+ \to 2_1^+)$ in Fig. 4.35b also presents a clear signature of the structural evolution involving triaxiality. For Yb, Hf, W, and Os isotopes with $110 \le N \le 116$, the branching ratio decreases from values close to the SU(3) limit of 0.7 to the U(5)/O(6) limit of zero. This behavior corresponds to the transition from well deformed nuclei to γ-soft as confirmed by the experimental data. At this point, one can observe the increase from $N = 116$ toward the shell closure $N = 126$. The increase represents the deviation from the γ softness, as the corresponding mapped energy surface in Fig. 4.29 exhibits weak oblate deformation. The change in the branching ratio occurs slowly compared with the D1S case [106]. This is consistent with our general finding that the D1M energy surfaces for these nuclei suggest the quadrupole correlation less pronounced than the D1S ones. The branching ratios for Pt isotopes remain always much closer to zero in Fig. 4.35b, which is compatible with their γ-soft character.

Brief Summary

To summarize Sect. 4.3.3, we have analyzed the low-lying collective structure of the neutron-rich Yb, Hf, W, Os, and Pt isotopes in terms of the IBM-2 derived from the new parametrization Gogny D1M. The binding energy surface obtained from the constrained HFB calculation serves as a starting point for both reproducing and predicting the ground-state shape of the considered nuclei. The merits of the Gogny-HFB method and the IBM-2 framework have been exploited.

It was shown that Pt isotopes differ in the rapidity of the shape transition from other isotopes. The mapped IBM-2 energy surfaces for most of the considered Pt nuclei are γ unstable. The transition occurs more rapidly when departing from $Z = 76$ (Os) through $Z = 70$ (Yb). The triaxial deformation helps understand the prolate-to-oblate shape transition that occurs in the considered isotopes. The $N = 116$ nuclei can be commonly identified as the transition points. This is most noticeably seen from the systematics of the bandhead of the γ band 2_2^+. These features are confirmed through the analyses of spectra, ground-state correlation energy, and $B(E2)$ systematics. Predicted spectra have been presented for the neutron-rich Yb and Hf isotopes, where a quite rapid structural evolution is suggested.

Compared to the results from the standard parametrization Gogny-D1S [72, 106], it is likely that the D1M functional is equally valid. Further studies should be yet needed to probe which EDF is most appropriate for more precise mapping procedure.

Last, we here point out the following problems inherent to the prolate-oblate dynamics within IBM-2 and the possibilities to improve the current model:

- The discrepancies in the γ-band structure should not be overlooked. This is mainly due to the use of the IBM-2 Hamiltonian not reproducing the triaxial energy minimum. A specific three-body (cubic) term may improve the agreement. This problem has been partly answered in Ref. [125], and will be considered in Chap. 6.
- It is still to be clearified whether a single configuration without the particle-hole excitation holds for the considered mass region. Further studies would be needed

4.3 Prolate-Oblate Shape Dynamics

to compare the single and the mixed IBM configurations in a microscopic way, although the number of free parameters would increase drastically.
- In calculating the $B(E2)$ values, the boson effective charges cannot be obviously determined in a microscopic way. The most straightforward solution would be to equate the intrinsic quadrupole moment of the mean-field calculation to the intrinsic quadrupole moment of the IBM-2 system. Rather than this, however, since the effective charge is used for the purpose of spectroscopy, it is likely that one should take into account the effect beyond the mean field, namely the core polarization. This is an interesting future work, and is one of our on-going projects.
- Also, it would be meaningful to compare the spectra and the electromagnetic transition rates resulting from the present method in the region of $A \approx 190$ directly with those obtained from the full configuration-mixing and symmetry-conserving calculations that includes triaxial degrees of freedom, in order to quantify the predictive power of the employed model. Work along this direction has just started and is in progress.

4.3.4 QPT in Exotic Nuclei

The neutron-rich nuclei in the mass number $A \approx 100$ have been an ideal testing ground for the competition between the excitation modes originated from the single-particle and the collective degrees of freedom, and have indeed attracted considerable attention over the past four decades both experimentally and theoretically [126–130]. For instance, sudden onset of deformation has been observed in the even-even Zr ($Z = 40$) and Sr ($Z = 38$) isotopes when the neutron number N is changed from 58 to 60. For $N \leq 58$ the shapes for both isotopes are almost spherical, whereas strongly deformed shapes have been confirmed for $N \geq 60$ [127]. The mechanism behind these phenomena can be ascribed to the strongly interacting proton and neutron Nilsson orbits (cf. Ref. [130] and references therein). For neutrons, on the one hand, the downsloping of the $\nu_{1/2} - [505]$ and the $\nu_{3/2} - [541]$ orbits, which occurs due to the splitting of the spherical $\nu h_{11/2}$ orbital, gives rise to the deformation, whereas the extruder $\nu g_{9/2} + [404]$ orbital stabilizes the deformation at a saturation level of approximately $\beta \approx 0.4$ [130] with β being the axially symmetric deformation in the geometrical model. For protons, on the other hand, the downsloping of the $\pi_{1/2} + [440]$ and the $\pi_{3/2} + [431]$ orbits, originating from the spherical $\pi g_{9/2}$ orbital, are fully occupied at $Z = 38$ and 40 with the deformation of $\beta \approx 0.4$. At the neutron number $N = 60$, the deformed configuration becomes dominant over the spherical one, thereby resulting in the observed strong deformation.

In this context, when descending to the Krypton isotopes with proton number $Z = 36$, only the proton $\pi_{1/2} + [440]$ orbital is fully occupied with the deformation of $\beta \approx 0.4$. Therefore the question arises as to whether the reduced occupation of the deformation driving proton intruder orbitals is still strong enough to enhance deformation as rapidly as neighbouring Sr and Zr isotopes. It should be noted that the spectroscopic data for the Kr nuclei with $N \geq 60$ have been quite sparse. Even the most basic measure of 2_1^+ excitation energies have been known only up to ^{94}Kr ($N = 58$), and the $B(E2)$ value for the $2_1^+ \to 0_1^+$ transition is known only for ^{88}Kr and ^{92}Kr nuclei [131]. By the experimental study in 2009 [132], the γ ray with the energy of $E_\gamma = 241$ keV was assigned to the $2_1^+ \to 0_1^+$ transition for ^{96}Kr ($N = 60$) nucleus. In the measurement reported in Ref. [132], sudden change of the deformation from $N = 58$ to 60 was suggested, similar to the cases in Sr and Zr isotopes.

The finding of [132] has been denied by the mass measurement, using the ISOLTRAP Penning-trap spectrometer at CERN[3] ISOLDE[4] facility [133]. According to the measurement of [133], the two-neutron separation energies evolve smoothly toward $N = 60$ for Kr isotopes, implying the smooth onset of deformation at this neutron number, which is contrary to the results in Ref. [132]. It should be then quite interesting to see whether similarly smooth evolution of nuclear structure in Kr isotopes can be found in the spectroscopic observables. To clarify the contradiction between the measurements of Refs. [132] and [133], the experiments were performed to measure the $B(E2; 2_1^+ \to 0_1^+)$ values of the neutron-rich exotic ^{94,96}Kr nuclei, employing the technique of sub-barrier projectile Coulomb excitation [134]. The experiment was carried out at the REX-ISOLDE[5] facility at CERN. This experiment supported the conclusion of the mass measurement of [133], identifying the $E_\gamma = 554.1(5)$ keV as the $2_1^+ \to 0_1^+$ transition. The observed 2_1^+ excitation energies and the $B(E2)$ values of the neutron-rich Kr isotopes can be a first hint implying the smooth onset of deformation. In addition, this new experimental result was supported by the theoretical calculation employing the IBM-2 Hamiltonian determined based on the constrained HFB approach with the microscopic Gogny-D1M energy density functional. The D1M functional is taken because it is more oriented to the nuclear mass model [52]. Note that there is no notable difference between the cases of the D1M functional and the standard parametrization of D1S functional.

This section illustrates the work of Ref. [134], but focuses on the theoretical aspect and the resultant physical interpretation presented there. The reader who is interested in the experimental detail is referred to the original paper [134].

To begin with, the newly obtained experimental data for the 2_1^+ excitation energies, E2 transition strengths, the lifetimes extracted from the $B(E2)$ values, and the determined quadrupole moments for the considered ^{94,96}Kr nuclei are summarized in Table 4.2, and the systematic trends of the experimental 2_1^+ excitation energies and the absolute $B(E2; 2_1^+ \to 0_1^+)$ values for the Kr isotopic chains from $N = 50$ to 60 are shown in Fig. 4.36. The energy of the excited 2_1^+ state observed in Ref. [132] for the ^{96}Kr nucleus is shown also for comparison. Figure 4.36 obviously shows that the experimental 2_1^+ excitation energies change quite smoothly when one goes from the neutron number $N = 58$ to 60, implying the modest onset of deformation as function of the neutron number. This finding is quite consistent with what is expected from the recently published result of the mass measurement [133], as

---

[3] Conseil Européen pour la Recherche Nucléaire.

[4] Isotope Separator On Line Facility for Production of Radioactive Ion-Beams.

[5] Radioactive Beam EXperiment at ISOLDE.

4.3 Prolate-Oblate Shape Dynamics

Table 4.2 Obtained experimental data for the 2_1^+ excitation energies, E2 transition strengths, the lifetimes extracted from the $B(E2)$ values, and the determined quadrupole moments

Isotopes	$E_\gamma(2_1^+ \to 0_1^+)$ (keV)	$B(E2; 2_1^+ \to 0_1^+)$ (W.u.)	$\tau(2_1^+)$ (ps)	$Q_{2_1^+}$ (b)
^{94}Kr	666.1(3)	$19.5^{+2.2}_{-2.1}$	$12.5^{+1.5}_{-1.2}$	$-0.45^{+0.33}_{-0.30}$
^{96}Kr	554.1(5)	$33.4^{+7.4}_{-6.7}$	$17.9^{+4.5}_{-3.3}$	0.26(92)

Data are taken from Ref. [134]

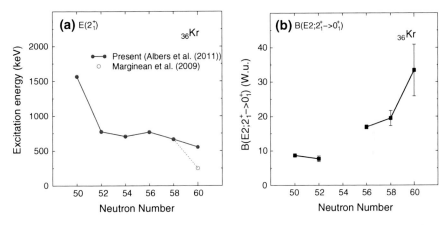

Fig. 4.36 The experimental data for the (**a**) 2_1^+ excitation energies and (**b**) the absolute $B(E2; 2_1^+ \to 0_1^+)$ values for the considered Kr isotopes. The *solid circles* connected with each other by the *solid curves* and the *open circle* in panel **a** represent the experimental data resulting from the present work [134] and from Ref. [132], respectively

well as with the results obtained from the isotopes shift δr_c^2 measurement of [135]. The latter was discussed theoretically in Ref. [136], where the smooth behavior in Kr isotopes was interpreted as a stabilization of the oblate shapes along the isotopic chain.

The assignment of $E_\gamma(2_1^+ \to 0_1^+) = 551$ keV γ ray can be supported by the IBM-2 calculation based on the constrained self-consistent HFB method with the D1M functional. It is sufficient to use the IBM-2 Hamiltonian of of the most standard form introduced in Eq. (2.41), consisting of only a few essential parts of a general two-body IBM-2 Hamiltonian, because one encounters no such situation as the stable triaxial minimum nor the coexisting minima in the microscopic Gogny EDF calculations for the considered Kr isotopes. Note that, for relatively light exotic nuclei, the conventional magic numbers may become no longer valid. We do not consider such effects and assume ^{78}Ni core as the boson vacuum. Thus, the proton boson number N_π is fixed, $N_\pi = 4$, and the neutron boson number N_ν changes from 0 to 5 for $^{86-96}$Kr nuclei, respectively. Also, since protons and neutrons occupy different major shells under this assumption, it is likely that one does not have to deal with the proton-neutron pairs and thus the usual IBM-2 framework suffices.

122　　　　　　　　　　　　　　4　Weakly Deformed Systems with Triaxial Dynamics

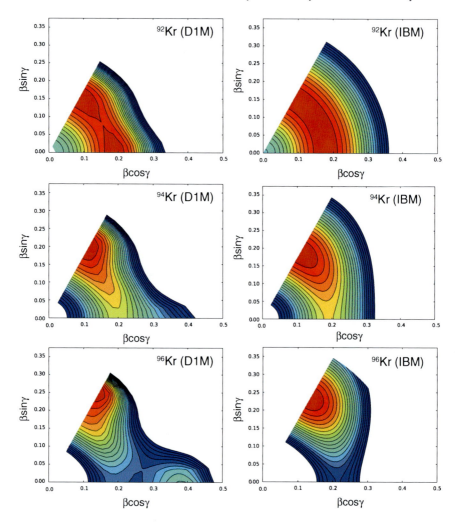

Fig. 4.37 The self-consistent energy landscapes with quadrupole degrees of freedom calculated by the constrained HFB method with Gogny-D1M EDF (*left panels*) and the mapped IBM-2 energy surfaces (*right panels*) for the $^{92-96}$Kr nuclei, drawn up to 2 MeV excitation in energy from the minima within the range of polar deformation parameter of $0° \leq \gamma \leq 60°$. The contour spacing is 100 keV

Figure 4.37 displays the self-consistent constrained energy surfaces with Gogny-D1M EDF and the mapped IBM-2 energy surfaces for the selected nuclei ^{92,94,96}Kr. The microscopic energy surface for ^{92}Kr appears to be totally flat and, for ^{94,96}Kr nuclei the oblate deformation is predicted. The potential becomes deeper from $N = 58$ to 60, but quite gradually. The mapped IBM energy surfaces indicate similar pattern. A local minimum for ^{96}Kr around $\beta = 0.4$ would be of little importance for the low-lying state and is not taken into account in the fit of the IBM-2 Hamiltonian.

4.3 Prolate-Oblate Shape Dynamics

Table 4.3 The parameters for the IBM-2 Hamiltonian obtained from the mapping of HFB to IBM-2 energy surfaces for the considered Kr nuclei with $50 \leq N \leq 60$

Nuclei	ε (MeV)	κ (MeV)	χ_π	χ_ν
^{86}Kr	1.3	−0.380	0.4	−0.4
^{88}Kr	0.947	−0.428	0.631	−0.533
^{90}Kr	1.052	−0.363	0.401	−0.438
^{92}Kr	1.211	−0.372	0.386	−0.392
^{94}Kr	1.208	−0.362	0.374	−0.020
^{96}Kr	1.280	−0.351	0.555	0.060

The derived IBM-2 parameters are listed in Table 4.3. What is worth noting is the overall systematic behaviour of the sum of the proton and neutron parameters $\chi_\pi + \chi_\nu$, reflecting totally flat collective potential energy surface for lighter isotopes being close to $\chi_\pi + \chi_\nu \approx 0$ and oblate configuration with $\chi_\pi + \chi_\nu > 0$ for heavier ones.

To calculate the absolute $B(E2)$ values, the E2 operator of the form in Eq. (2.50) is taken. The effective proton boson charge of $e_\pi = 0.07$ eb is fixed by adjusting to the experimental [137] $B(E2; 2_1^+ \to 0_1^+)$ value of the ^{86}Kr nucleus. In order to reduce the number of free parameters, the effective neutron boson charge of $e_\nu = 0.0$ eb is taken, following the idea of Refs. [138, 139].

The level schemes of all the considered Kr nuclei with $50 \leq N \leq 60$ are shown in Fig. 4.38, where the theoretical and the experimental 2_1^+, 4_1^+, 2_2^+ and 0_2^+ excitation energies as well as the absolute $B(E2; 2_1^+ \to 0_1^+)$ values are compared. The calculated 2_1^+ excitation energy of 534 keV for the exotic ^{96}Kr nucleus agrees quite well with the new experimental value of 554.1 keV. The present calculation predicts that the energies of the 0_2^+ and the 2_2^+ side-band states become larger than and

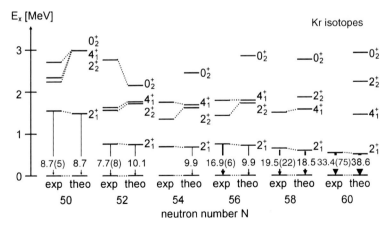

Fig. 4.38 Experimental and theoretical 2_1^+, 4_1^+, 2_2^+ and 0_2^+ excitation energies and the absolute $B(E2; 2_1^+ \to 0_1^+)$ values (in Weisskopf unit) for the considered Kr nuclei with $50 \leq N \leq 60$. The experimental $B(E2; 2_1^+ \to 0_1^+)$ values for ^{86,88,92}Kr nuclei are taken from Refs. [131, 137]. Figure is base on Ref. [134]

deviate from the 4_1^+ level from $N=56$ to 58 and to 60, implying the structural evolution to more deformed states. Good agreement is obtained also for the absolute $B(\text{E2}; 2_1^+ \to 0_1^+)$ values, except for the transitional nucleus ^{92}Kr. What is of particular significance should be that the present calculation seems to support the observed gradual evolution of deformation from $N=58$ to 60.

To summarize Sect. 4.3.4, new experimental results on the 2_1^+ excitation energies and the absolute $B(\text{E2}; 2_1^+ \to 0_1^+)$ values for the neutron-rich ^{94,96}Kr have been analyzed theoretically. These experimental findings reveal that, contrary to what was concluded in Ref. [132], no sudden onset of deformation is observed from $N=58$ to 60. The experiment supports, for the first time, the mass measurement of [133] from the spectroscopic point of view. The physical interpretation suggested by the new experiment has been confirmed in terms of the IBM-2 Hamiltonian determined based on the constrained HFB approach using the microscopic Gogny-D1M energy density functional.

References

1. Casten RF (2006) Shape phase transitions and critical-point phenomena in atomic nuclei. Nat Phys 2:811
2. Cejnar P, Casten RF, Jolie J (2010) Quantum phase transitions in the shapes of atomic nuclei. Rev Mod Phys 82:2155
3. Iachello F (2001) Analytic description of critical point nuclei in a spherical-axially deformed shape phase transition. Phys Rev Lett 87:052501
4. Iachello F (2000) Dynamical symmetries at the critical point. Phys Rev Lett 85:3580
5. Casten RF, Zamfir NV (2001) Empirical realization of a critical point description in atomic nuclei. Phys Rev Lett 87:052503
6. Casten RF, Zamfir NV (2000) evidence for a possible E(5) symmetry in ^{134}Ba. Phys Rev Lett 85:3584
7. Jolie J, Casten RF, von Brentano P, Werner V (2001) Quantum phase transition for γ-soft nuclei. Phys Rev Lett 87:16501
8. Kotila J, Nomura K, Guo L, Shimizu N, Otsuka T (2012) Shape phase transitions in the interacting boson model: phenomenological versus microscopic descriptions. Phys Rev C 85:054302
9. Bonche P, Flocard H, Heenen PH (2005) Solution of the skyrme HF + BCS equation on a 3D mesh. Comput Phys Commun 171:49
10. Bartel J, Quentin Ph, Brack M, Guet C, Håkansson H-B (1982) Towards a better parametrisation of skyrme-like effective forces: a critical study of the SkM force. Nucl Phys A 386:79
11. Nomura K, Shimizu N, Otsuka T (2010) Formulating the interacting Boson model by mean-field methods. Phys Rev C 81:044307
12. Brookhaven National Nuclear Data Center (NNDC). http://www.nndc.bnl.gov/nudat2/index.jsp
13. Van Isacker P, Puddu G (1980) The Ru and Pd isotopes in the proton-neutron interacting boson model. Nucl Phys A 348:125
14. Van Isacker P, Chen J-Q (1981) Classical limit of the interacting boson Hamiltonian. Phys Rev C 24:684
15. Heyde K, Van Isacker P, Waroquier M, Moreau J (1984) Triaxial shapes in the interacting boson model. Phys Rev C29:1420
16. Casten RF, von Brentano P, Heyde K, Van Isacker P, Jolie J (1985) The interplay of γ-softness and triaxiality in O(6)-like nuclei. Nucl Phys A 439:289

17. Stefanescu I, Gelberg A, Jolie J, Van Isacker P, von Brentano P, Luo YX, Zhu SJ, Rasmussen JO, Hamilton JH, Ramayya AV, Che XL (2007) IBM-1 description of the fission products ^{108,110,112}Ru. Nucl Phys A 789:125
18. Sorgunlu B, Van Isacker P (2008) Triaxiality in the interacting boson model. Nucl Phys A808:27–46
19. Williams E, Plettner C, McCutchan EA, Levine H, Zamfir NV, Cakirli RB, Casten RF, Ai H, Beausang CW, Gürdal G, Heinz A, Qian J, Meyer DA, Pietralla N, Werner V (2006) Revisiting anomalous $B(E2; 4_1^+ \to 2_1^+)/B(E2; 2_1^+ \to 0_1^+)$ values in ^{98}Ru and ^{180}Pt. Phys Rev C 74:024302
20. Srebrby J, Czosnyka T, Droste Ch, Rohoziński SG, Próchniak L, Zając K, Pomorski K, Cline D, Wu CY, Bäcklin A, Hasselgren L, Diamond RM, Habs D, Körner HJ, Stephens FS, Baktash C, Kostecki RP (2006) Experimental and theoretical investigations of quadrupole collective degrees of freedom in ^{104}Ru. Nucl Phys A 766:25
21. Singh B (2008) Nuclear data sheets for A = 100. Nucl Data Sheets 109:297
22. De Frenne D (2009) Nuclear data sheets for A = 102. Nucl Data Sheets 110:1745
23. Blanchot J (2007) Nuclear data sheets for A = 104. Nucl Data Sheets 108:2035
24. De Frenne D, Negret A (2008) Nuclear data sheets for A = 106. Nucl Data Sheets 109:943
25. Blachot J (2000) Nuclear data sheets for A = 108. Nucl Data Sheets 91:135
26. De Frenne D, Jacobs E (2000) Nuclear data sheets for A = 110. Nucl Data Sheets 89:481
27. Puddu G, Scholten O, Otsuka T (1980) Collective quadrupole states of Xe, Ba and Ce in the interacting boson model. Nucl Phys 348:109–124
28. Davydov AS, Filippov GF (1958) Rotational states in even atomic nuclei. Nucl Phys 8:237
29. Mukhopadhyay S, Scheck M, Crider B, Choudry SN, Elhami E, Peters E, McEllistrem MT, Orce1 JN, Yates SW, (2008) Multiphonon states in $^{136}_{56}$Ba$_{80}$. Phys Rev C 78:034317
30. Ahn T, Coquard L, Pietralla N, Rainovski G, Costin A, Janssens RVF, Lister CJ, Carpenter M, Zhu S, Heyde K (2009) Evolution of the one-phonon $2_{1,\text{ms}}^+$ mixed-symmetry state in $N = 80$ isotones as a local measure for the proton–neutron quadrupole interaction. Phys Lett B 679:19
31. Gade A, Wiedenhöver I, Meise H, Gelberg A, von Brentano P (2002) Decay properties of low-lying collective states in ^{132}Ba. Nucl Phys A 697:75
32. Coquard L, Pietralla N, Ahn T, Rainovski G, Bettermann L, Carpenter MP, Janssens RVF, Leske J, Lister CJ, Möller O, Rother W, Werner V, Zhu S (2009) Robust test of E(5) symmetry in ^{128}Xe. Phys Rev C 80:061304(R)
33. Pascu S, Cata-Danil Gh, Bucurescu D, Marginean N, Zamfir NV, Graw G, Gollwitzer A, Hofer D, Valnion BD (2009) Investigation of the ^{128}Ba nucleus with the (p, t) reaction. Phys Rev C 79:064323
34. Sonzogni AA (2004) Nuclear data sheets for A = 134. Nucl Data Sheets 103:1
35. Khazov Yu, Rodionov AA, Sakharov S, Singh B (2005) Nuclear data sheets for A = 132. Nucl Data Sheets 104:497
36. Kanbe M, Kitao K (2001) Nuclear data sheets for A = 128. Nucl Data Sheets 94:227
37. Katakura J, Wu ZD (2008) Nuclear data sheets for A = 124. Nucl Data Sheets 109:1655
38. Tamura T (2007) Nuclear data sheets for A = 122. Nucl Data Sheets 108:455
39. Kitao K, Tendow Y, Hashizume A (2002) Nuclear Data Sheets for A = 120. Nucl Data Sheets 96:241
40. Kitao K (1995) Nuclear data sheets for A = 118. Nucl Data Sheets 75:99
41. Otsuka T, Arima A, Iachello F, Talmi I (1978) Shell model description of interacting bosons. Phys Lett B 76:139
42. Otsuka T, Arima A, Iachello F (1978) Shell model description of interacting bosons. Nucl Phys A 309:1
43. Otsuka T (1993) Microscopic basis and introduction to IBM-2. In: Casten RF (ed) Algebraic approaches to nuclear structure. Harwood, Chur, p 195
44. Sarriguren P, Rodríguez-Guzmán R, Robledo LM (2008) Shape transitions in neutron-rich Yb, Hf, W, Os, and Pt isotopes within a Skyrme Hartree-Fock + BCS approach. Phys Rev C 77:064322

45. Egido JL, Robledo LM, Rodríguez-Guzmán RR (2004) Unveiling the origin of shape coexistence in lead isotopes. Phys Rev Lett 93:082502
46. Robledo LM, Rodríguez-Guzmán R, Sarriguren P (2009) Role of triaxiality in the ground-state shape of neutron-rich Yb, Hf, W, Os and Pt isotopes. J Phys G: Nucl Part Phys 36:115104
47. Berger JF, Girod M, Gogny D (1984) Microscopic analysis of collective dynamics in low energy fission. Nucl Phys A 428:23c
48. Decharge J, Girod M, Gogny D (1975) Self consistent calculations and quadrupole moments of even Sm isotopes. Phys Lett B 55:361
49. Dechargé J, Gogny D (1980) Hartree-Fock-Bogolyubov calculations with the D1 effective interaction on spherical nuclei. Phys Rev C 21:1568
50. Rodriguez-Guzman R, Sarriguren P, Robledo LM, Garcis-Ramos JE (2010) Mean field study of structural changes in Pt isotopes with the Gogny interaction. Phys Rev C 81:024310
51. Chappert F, Girod M, Hilaire S (2008) Towards a new Gogny force parametrization: impact of the neutron matter equation of state. Phys Lett B 668:420
52. Goriely S, Hilaire S, Girod M, Peru S (2009) First Gogny-Hartree-Fock-Bogoliubov nuclear mass model. Phys Rev Lett 102:242501
53. Nomura K, Shimizu N, Otsuka T (2008) Mean-field derivation of the interacting Boson model Hamiltonian and exotic nuclei. Phys Rev Lett 101:142501
54. Delaroche J-P, Girod M, Libert L, Goutte H, Hilaire S, Peru S, Pillet N, Bertsch GF (2010) Structure of even-even nuclei using a mapped collective Hamiltonian and the D1S Gogny interaction. Phys Rev C 81:014303
55. Cizewski JA, Casten RF, Smith GJ, Stelts ML, Kane WR, Borner HG, Davidson WF (1978) Evidence for a new symmetry in nuclei: the structure of ^{196}Pt and the O(6) limit. Phys Rev Lett 40:167
56. Casten RF, Cizewski JA (1978) The O(6)→rotor transition in the Pt-Os nuclei. Nucl Phys A 309:477
57. Duval P, Barrett BR (1981) Interacting boson approximation model of the tungsten isotopes. Phys Rev C 23:492
58. Bijker R, Dieperink AEL, Scholten O, Spanhoff R (1980) Description of the Pt and Os isotopes in the interacting boson model. Nucl Phys A 344:207
59. Podolyák Zs et al (2009) Weakly deformed oblate structures in $^{198}_{76}$Os$_{122}$. Phys Rev C 79:031305(R)
60. Kummer K, Baranger M (1968) Nuclear deformations in the pairing-plus-quadrupole model (V). Energy levels and electromagnetic moments of the W, Os and Pt nuclei. Nucl Phys A122:273
61. Delaroche J-P, Girod M, Bastin G, Deloncle I, Hannachi F, Libert J, Porquet MG, Bourgeois C, Hojman D, Kilcher P, Korichi A, Le Blanc F, Perrin N, Roussiere B, Sauvage J, Sergolle H (1994) Evidence for γ vibrations and shape evolutions through the transitional ^{184,186,188,190}Hg nuclei. Phys Rev C 50:2332
62. Nikšić T, Ring P, Vretenar D, Tian T, Ma Z (2010) 3D relativistic Hartree-Bogoliubov model with a separable pairing interaction: triaxial ground-state shapes. Phys Rev C 81:054318
63. Andreyev AN et al (2005) A triplet of differently shaped spin-zero states in the atomic nucleus ^{186}Pb. Nature (London) 405:430
64. Rodriguez-Guzman RR, Egido JL, Robledo LM (2004) Beyond mean field description of shape coexistence in neutron-deficient Pb isotopes. Phys Rev C 69:054319
65. Duguet T, Bender M, Bonche P, Heenen P-H (2003) Shape coexistence in ^{186}Pb: beyond-mean-field description by configuration mixing of symmetry restored wave functions. Phys Lett B 559:201
66. Bender M, Bonche P, Duguet T, Heenen P-H (2004) Configuration mixing of angular momentum projected self-consistent mean-field states for neutron-deficient Pb isotopes. Phys Rev C 69:064303
67. Duval PD, Barrett BR (1983) Quantitative description of configuration mixing in the interacting boson model. Nucl Phys A 376:213

References

68. McCutchan EA, Casten RF, Zamfir V (2005) Simple interpretation of shape evolution in Pt isotopes without intruder states. Phys Rev C 71:061301(R)
69. Garcia-Ramos JE, Heyde K (2009) The Pt isotopes: comparing the interacting boson model with configuration mixing and the extended consistent-Q formalism. Nucl Phys A 825:39
70. Egido JL, Lessing J, Martin V, Robledo LM (1995) On the solution of the Hartree-Fock-Bogoliubov equations by the conjugate gradient method. Nucl Phys A 594:70
71. Ring P, Schuck P (1980) The nuclear many-body problem. Springer, Berlin
72. Nomura K, Otsuka T, Rodríguez-Guzmán R, Robledo LM, Sarriguren P (2011) Structural evolution in Pt isotopes with the interacting boson model Hamiltonian derived from the Gogny energy density functional. Phys Rev C 83:014309
73. Bohr A, Mottelson BR (1969) Nuclear structure, vol I. Single-particle motion. Benjamin, New York
74. Bohr A, Mottelson BR (1975) Nuclear structure, vol II. Nuclear deformations. Benjamin, New York
75. Ginocchio JN, Kirson M (1980) An intrinsic state for the interacting boson model and its relationship to the Bohr-Mottelson model. Nucl Phys A 350:31
76. Baglin CM (2002) Nuclear data sheets for A = 170. Nucl Data Sheets 96:611
77. Balraj S (1995) Nuclear data sheets for A = 172. Nucl Data Sheets 75:199
78. Browne E, Junde H (1999) Nuclear data sheets for A = 174. Nucl Data Sheets 87:15
79. Basunia MS (2006) Nuclear data sheets for A = 176. Nucl Data Sheets 107:791
80. Achterberg E et al (2009) Nuclear data sheets for A = 178. Nucl Data Sheets 110:1473
81. Wu S-C, Niu H (2003) Nuclear data sheets for A = 180. Nucl Data Sheets 100:483
82. Singh B, Firestone RB (1995) Nuclear data sheets for A = 182. Nucl Data Sheets 74:383
83. Baglin CM (2010) Nuclear data sheets for A = 184. Nucl Data Sheets 111:275
84. Baglin CM (2003) Nuclear data sheets for A = 186. Nucl Data Sheets 99:1
85. Singh B (2002) Nuclear data sheets for A = 188. Nucl Data Sheets 95:387
86. Singh B (2003) Nuclear data sheets for A = 190. Nucl Data Sheets 99:275
87. Baglin CM (1998) Nuclear data sheets for A = 192. Nucl Data Sheets 84:717
88. Singh B (2006) Nuclear data sheets for A = 194. Nucl Data Sheets 107:1531
89. Xiaolong H (2007) Nuclear data sheets for A = 196. Nucl Data Sheets 108:1093
90. Xiaolong H (2007) Nuclear data sheets for A = 198. Nucl Data Sheets 110:2533
91. Kondeva FG, Lalkovski S (2007) Nuclear data sheets for A = 200. Nucl Data Sheets 108:1471
92. Otsuka T, Yoshida N (1985) User's manual of the program NPBOS. JAERI-M report 85, Japan Atomic Energy Research Institute
93. Arima A, Iachello F (1975) Collective nuclear states as representations of a SU(6) group. Phys Rev Lett 35:1069
94. Iachello F, Arima A (1987) The interacting boson model. Cambridge University Press, Cambridge
95. Scholten O, Iachello F, Arima A (1978) Interacting boson model of collective nuclear states III. The transition from SU(5) to SU(3). Ann Phys (NY) 115:325
96. Arima A, Iachello F (1979) Interacting boson model of collective states: IV. The O(6) limit. Ann Phys 123:468–492
97. Shizuma T, Ishii T, Makii H, Hayakawa T, Shigematsu S, Matsuda M, Ideguchi E, Zheng Y, Liu M, Morikawa T, Walker PM, Oi M (2006) Excited states in neutron-rich ^{188}W produced by an ^{18}O-induced 2-neutron transfer reaction. Eur Phys J A 30:391
98. Zs Podolyák et al (2000) Isomer spectroscopy of neutron rich ^{190}W$_{116}$. Phys Lett B 491:225
99. Jolie J, Linnemann A (2003) Prolate-oblate phase transition in the Hf-Hg mass region. Phys Rev C 68:031301(R)
100. Alkhomashi N et al (2009) β^--delayed spectroscopy of neutron-rich tantalum nuclei: shape evolution in neutron-rich tungsten isotopes. Phys Rev C 80:064308
101. Regan PH et al (2008) First results with the rising active stopper. Int J Mod Phys E17(Suppl 1):8

102. Lane GJ, Dracoulis GD, Kondev FG, Hughes RO, Watanabe H, Byrne AP, Carpenter MP, Chiara CJ, Chowdhury P, Janssens RVF, Lauritsen T, Lister CJ, McCutchan EA, Seweryniak D, Stefanescu I, Zhu S (2010) Structure of neutron-rich tungsten nuclei and evidence for a 10^- isomer in ^{190}W. Phys Rev C 82:051304
103. Wheldon C, Garcés Narro J, Pearson CJ, Regan PH, Zs Podolyák, Warner DD, Fallon P, Macchiavelli AO, Cromaz M (2001) Yrast states in ^{194}Os: the prolate-oblate transition region. Phys Rev C 63:011304
104. Bond PD, Casten RF, Warner DD, Horn D (1983) Excited states in 1960s and the structure of the Pt-Os transition region. Phys Lett B 130:167
105. Stevenson PD, Brine MP, Zs Podolyák, Regan PH, Walker PM, Rikovska Stone J (2005) Shape evolution in the neutron-rich tungsten region. Phys Rev C 72:047303
106. Nomura K, Otsuka T, Rodríguez-Guzmán R, Robledo LM, Sarriguren P, Regan PH, Stevenson PD, Zs Podolyák (2011) Spectroscopic calculations of the low-lying structure in exotic Os and W isotopes. Phys Rev C 83:051303
107. Rodríguez-Guzmán R, Sarriguren P, Robledo LM, Perez-Martín S (2000) Charge radii and structural evolution in Sr, Zr, and Mo isotopes. Phys Lett B 691:202
108. Rodríguez-Guzmán R, Sarriguren P, Robledo LM (2010) Systematics of one-quasiparticle configurations in neutron-rich odd Sr, Zr, and Mo isotopes with the Gogny energy density functional. Phys Rev C 82:044318
109. Rodríguez-Guzmán R, Sarriguren P, Robledo LM (2010) Signatures of shape transitions in odd-A neutron-rich rubidium isotopes. Phys Rev C 82:061302(R)
110. Nomura K, Otsuka T, Shimizu N, Guo L (2011) Microscopic formulation of the interacting Boson model for rotational nuclei. Phys Rev C 83:041302(R)
111. Thouless DJ, Valatin JG (1962) Time-dependent Hartree-Fock equations and rotational states of nuclei. Nucl Phys 31:211
112. Inglis DR (1956) Nuclear moments of inertia due to nucleon motion in a rotating well. Phys Rev 103:1786
113. Belyaev ST (1961) Concerning the calculation of the nuclear moment of inertia. Nucl Phys 24:322
114. Schaaser H, Brink DM (1984) calculations away from SU(3) symmetry by cranking the interacting boson model. Phys Lett B 143:269
115. Nomura K, Otsuka T, Rodríguez-Guzmán R, Robledo LM, Sarriguren P (2011) Collective structural evolution in Yb, Hf, W, Os and Pt isotopes. Phys Rev C 84:054316
116. Bender M, Bertsch G, Heenen P-H (2006) Global study of quadrupole correlation effects. Phys Rev C 73:034322
117. Yao JM, Mei H, Chen H, Meng J, Ring P, Vretenar D (2011) Configuration mixing of angular-momentum-projected triaxial relativistic mean-field wave functions. II. Microscopic analysis of low-lying states in magnesium isotopes. Phys Rev C 83:014308
118. Davidson PM, Dracoulis GD, Kibédi T, Byrne AP, Anderssen SS, Baxter AM, Fabricius B, Lane GJ, Stuchbery AE (1994) Non-yrast states and shape co-existence in ^{172}Os. Nucl Phys A 568:90
119. Davidson PM, Dracoulis GD, Kibédi T, Byrne AP, Anderssen SS, Baxter AM, Fabricius B, Lane GJ, Stuchbery AE (1999) Non-yrast states and shape co-existence in light Pt isotopes. Nucl Phys A 657:219
120. Kibéti T, Dracoulis GD, Byrne AP, Davidson PM (1994) Low-spin non-yrast states and collective excitations in ^{174}Os, ^{176}Os, ^{178}Os, ^{180}Os, ^{182}Os and ^{184}Os. Nucl Phys A 567:183
121. Kibéti T, Dracoulis GD, Byrne AP, Davidson PM (2001) Low-spin non-yrast states in light tungsten isotopes and the evolution of shape coexistence. Nucl Phys A 688:669
122. Wilets L, Jean M (1956) Surface oscillations in even-even nuclei. Phys Rev 102:788
123. Caprio MA, Iachello F (2004) Phase structure of the two-fluid proton-neutron system. Phys Rev Lett 93:242502
124. Caprio MA, Iachello F (2005) Phase structure of a two-fluid bosonic system. Ann Phys 318:454

125. Nomura K, Shimizu N, Vretenar D, Nikšić T, Otsuka T (2012) Robust regularity in γ-soft nuclei and its novel microscopic realization. Phys Rev Lett 108:132501
126. Federmann P, Pittel S (1977) Towards a unified microscopic description of nuclear deformation. Phys Lett B 69:385
127. Eberth J, Meyer RA, Sistemich K (eds) (1988) Nuclear structure of the Zr region. Springer, Berlin
128. Lhersonneau G, Pfeiffer B, Kratz K-L, Ohm H, Sistemich K (1988) Shape coexistence in the $N = 59$ isotone ^{97}Sr. Z Phys A 330:347
129. Hotchkis MAC, Durell JL, Fitzgerald JB, Mowbray AS, Phillips WR, Ahmad I, Carpenter MP, Janssens RVF, Khoo TL, Moore EF, Morss LR, Benet Ph, Ye D (1990) New neutron-rich nuclei ^{103,104}Zr and the $A \sim 100$ region of deformation. Phys Rev Lett 64:3123
130. Urban W, Pinston JA, Genevey J, Rzaça-Urban T, Złomaniec A, Simpson G, Durell JL, Phillips WR, Smith AG, Varley BJ, Ahmad I, Schulz N (2004) The ν9/2[404] orbital and the deformation in the $A \sim 100$ region. Eur Phys J A 22:241
131. Mücher D (2009) Dynamische symmetrien von atomkernen an unterschalenabschlüssen. PhD thesis, Insitutfür Kernphysik, Universität zu Köln. http://kups.ub.uni-koeln.de/2868/
132. Mărginean N, Bucurescu D, Ur CA, Mihai C, Corradi L, Farnea E, Filipescu D, Fioretto E, Ghiţă D, Guiot B, Górska M, Ionescu-Bujor M, Iordăchescu A, Jelavić-Malenica D, Lenzi SM, Mason P, Mărginean R, Mengoni D, Montagnoli G, Napoli DR, Pascu S, Pollarolo G, Recchia F, Stefanini AM, Silvestri R, Sava T, Scarlassara F, Szilner S, Zamfir NV (2009) Evolution of deformation in the neutron-rich krypton isotopes: the ^{96}Kr nucleus. Phys Rev C 80:021301(R)
133. Naimi S, Audi G, Beck D, Blaum K, Böhm Ch, Borgmann Ch, Breitenfeldt M, George S, Herfurth F, Herlert A, Kowalska M, Kreim S, Lunney D, Neidherr D, Rosenbusch M, Schwarz S, Schweikhard L, Zuber K (2010) Critical-point boundary for the nuclear quantum phase transition near $A = 100$ from mass measurements of ^{96,97}Kr. Phys Rev Lett 105:032502
134. Albers M, Warr N, Nomura K, Blazhev A, Jolie J, Mücher D, Bastin B, Bauer C, Bernards C, Bettermann L, Bildstein V, Butterworth J, Cappellazzo M, Cederkäll J, Cline D, Darby I, Daugas JM, Davinson T, De Witte H, Diriken J, Filipescu D, Fiori E, Fransen C, Gaffney LP, Georgiev G, Gernhäuser R, Hackstein M, Heinze S, Hess H, Huyse M, Jenkins D, Konki J, Kowalczyk M, Kröll T, Lutter R, Marginean N, Mihai C, Moschner K, Napiorkowski P, Nowak K, Otsuka T, Pakarinen J, Pfeiffer M, Radeck D, Reiter P, Rigby S, Robledo LM, Rodríguez-Guzmán R, Rudigier M, Sarriguren P, Scheck M, Seidlitz M, Siebeck B, Simpson G, Thoele P, Thomas T, Van de Walle J, Van Duppen P, Vermeulen M, Voulot D, Wadsworth R, Wenander F, Wimmer K, Zell KO, Zielinska M (2012) Evidence for a smooth onset of deformation in the neutron-rich Kr isotopes. Phys Rev Lett 108:062701
135. Keim M, Arnold E, Borchers W, Georg U, Klein A, Neugart R, Vermeeren L, Silverans RE, Lievens P (1995) Laser-spectroscopy measurements of $^{72-96}$Kr spins, moments and charge radii. Nucl Phys A 586:219
136. Rodríguez-Guzmán R, Sarriguren P, Robledo LM (2011) Shape evolution in yttrium and niobium neutron-rich isotopes. Phys Rev C 83:044307
137. Mertzimekis TJ, Benczer-Koller N, Holden J, Jakob G, Kumbartzki G, Speidel K-H, Ernst R, Macchiavelli A, McMahan M, Phair L, Maier-Komor P, Pakou A, Vincent S, Korten W (2001) First measurements of g factors in the even Kr isotopes. Phys Rev C 64:024314
138. Pietralla N, Belic D, von Brentano P, Fransen C, Herzberg R-D, Kneissl U, Maser H, Matschinsky P, Nord A, Otsuka T, Pitz HH, Werner V, Wiedenhöver I (1998) Isovector quadrupole excitations in the valence shell of the vibrator nucleus ^{136}Ba: evidence from photon scattering experiments. Phys Rev C 58:796
139. Fransen C, Pietralla N, Ammar Z, Bandyopadhyay D, Boukharouba N, von Brentano P, Dewald A, Gableske J, Gade A, Jolie J, Kneissl U, Lesher SR, Lisetskiy AF, McEllistrem MT, Merrick M, Pitz HH, Warr N, Werner V, Yates SW (2003) Comprehensive studies of low-spin collective excitations in ^{94}Mo. Phys Rev C 67:024307

Chapter 5
Comparison with Geometrical Model

5.1 Aim

From the practical side, the methodology described so far yields spectroscopic observables in a computationally moderate way, which can be an alternative to the full GCM configuration-mixing approach. In another sound approximation to the full GCM approach to five-dimensional quadrupole dynamics that restores rotational symmetry and allows for fluctuations around the triaxial mean-field minima, a collective Hamiltonian can be formulated, with deformation-dependent parameters determined by constrained microscopic self-consistent mean-field calculations. The dynamics of the five-dimensional Hamiltonian for quadrupole vibrational and rotational degrees of freedom is governed by the seven functions of the intrinsic quadrupole deformations: the collective potential, three vibrational mass parameters, and three moments of inertia for rotations around the principal axes [4–8].

It would be, therefore, interesting to compare the collective model and the IBM, starting from the same self-consistent mean-field solution based on a microscopic EDF. In this chapter, spectroscopic observables calculated with the IBM Hamiltonian are compared to the solution of the collective quadrupole Hamiltonian, with both calculations being based on relativistic Hartree-Bogoliubov (RHB) [9] self-consistent mean-field energy surfaces. The framework of relativistic EDFs and the corresponding collective Hamiltonian have successfully been employed in studies of the evolution of ground-state shapes and spectroscopic properties of medium-heavy and heavy nuclei [6–11]. In the present analysis we consider the even-even isotopes $^{192-196}$Pt. In the IBM framework these γ-soft nuclei can be characterized by the O(6) dynamical symmetry [12–15].

5.2 Bohr Hamiltonian

The map of the energy surface as a function of the quadrupole collective variables β and γ [16] is obtained from self-consistent RHB calculations with additional constraints on the axial and triaxial mass quadrupole moments. The quadrupole moments can be related to the polar deformation parameters β and γ. The parameter β is simply proportional to the intrinsic quadrupole moment, and the angular variable γ specifies the type and orientation of the shape. The limit $\gamma = 0$ corresponds to axial prolate shapes, whereas the shape is oblate for $\gamma = \pi/3$. Triaxial shapes are associated with intermediate values $0 < \gamma < \pi/3$. In this work the constrained RHB calculations have been performed using the relativistic functional DD-PC1 [18]. Starting from microscopic nucleon self-energies in nuclear matter, and empirical global properties of the nuclear matter equation of state, the coupling parameters of DD-PC1 have been determined in a careful comparison of the calculated binding energies with data, for a set of 64 axially deformed nuclei in the mass regions $A \approx 150–180$ and $A \approx 230–250$. DD-PC1 has been further tested in a series of calculations of properties of spherical and deformed medium-heavy and heavy nuclei, including binding energies, charge radii, deformation parameters, neutron skin thickness, and excitation energies of giant multipole resonances. For the examples presented here, pairing correlations have been taken into account by employing a pairing force that is separable in momentum space, and is completely determined by two parameters adjusted to reproduce the empirical bell-shaped pairing gap in symmetric nuclear matter [19].

The entire dynamics of the collective Hamiltonian is governed by seven functions of the intrinsic deformations β and γ: the collective potential, the three mass parameters: $B_{\beta\beta}$, $B_{\beta\gamma}$, $B_{\gamma\gamma}$, and the three moments of inertia $\mathscr{I}_k$. These functions are determined by the choice of a particular microscopic nuclear energy density functional and a pairing functional. The quasiparticle wave functions and energies, that correspond to constrained self-consistent solutions of the RHB model, provide the microscopic input for the parameters of the collective Hamiltonian [6]:

$$\hat{H}_{\text{coll}} = \hat{T}_{\text{vib}} + \hat{T}_{\text{rot}} + V_{\text{coll}}, \tag{5.1}$$

with the vibrational kinetic energy:

$$\hat{T}_{\text{vib}} = -\frac{\hbar^2}{2\sqrt{wr}} \left\{ \frac{1}{\beta^4} \left[\frac{\partial}{\partial \beta} \sqrt{\frac{r}{w}} \beta^4 B_{\gamma\gamma} \frac{\partial}{\partial \beta} - \frac{\partial}{\partial \beta} \sqrt{\frac{r}{w}} \beta^3 B_{\beta\gamma} \frac{\partial}{\partial \gamma} \right] \right.$$
$$\left. + \frac{1}{\beta \sin 3\gamma} \left[-\frac{\partial}{\partial \gamma} \sqrt{\frac{r}{w}} \sin 3\gamma \times B_{\beta\gamma} \frac{\partial}{\partial \beta} + \frac{1}{\beta} \frac{\partial}{\partial \gamma} \sqrt{\frac{r}{w}} \sin 3\gamma B_{\beta\beta} \frac{\partial}{\partial \gamma} \right] \right\}, \tag{5.2}$$

5.2 Bohr Hamiltonian

and rotational kinetic energy:

$$\hat{T}_{\text{rot}} = \frac{1}{2} \sum_{k=1}^{3} \frac{\hat{J}_k^2}{\mathscr{I}_k}. \tag{5.3}$$

V_{coll} is the collective potential. $\hat{J}_k$ denotes the components of the angular momentum in the body-fixed frame of a nucleus, and the mass parameters $B_{\beta\beta}$, $B_{\beta\gamma}$, $B_{\gamma\gamma}$, as well as the moments of inertia $\mathscr{I}_k$, depend on the quadrupole deformation variables β and γ: $\mathscr{I}_k = 4B_k\beta^2 \sin^2(\gamma - 2k\pi/3)$. Two additional quantities that appear in the expression for the vibrational energy: $r = B_1 B_2 B_3$, and $w = B_{\beta\beta}B_{\gamma\gamma} - B_{\beta\gamma}^2$, determine the volume element in the collective space. The moments of inertia are computed using the Inglis-Belyaev (IB) formula [20, 21], and the mass parameters associated with the two quadrupole collective coordinates $q_0 = \langle \hat{Q}_{20} \rangle$ and $q_2 = \langle \hat{Q}_{22} \rangle$ are calculated in the cranking approximation. The potential V_{coll} in the collective Hamiltonian Eq. (5.1) is obtained by subtracting the zero-point energy corrections from the total energy that corresponds to the solution of constrained RHB equations, at each point on the triaxial deformation plane.

$$V_{\text{coll}}(\beta, \gamma) = E_{\text{RMF}}(\beta, \gamma) - \Delta V_{\text{coll}}(\beta, \gamma) - \Delta V_{\text{rot}}(\beta, \gamma). \tag{5.4}$$

The Hamiltonian Eq. (5.1) describes quadrupole vibrations, rotations, and the coupling of these collective modes. The corresponding eigenvalue problem is solved using an expansion of eigenfunctions in terms of a complete set of basis functions that depend on the deformation variables β and γ, and the Euler angles ϕ, θ and ψ [6]. The diagonalization of the Hamiltonian yields the excitation energies and collective wave functions for each value of the total angular momentum and parity, that are used to calculate observables. An important advantage of using the collective model based on self-consistent mean-field single-(quasi)particle solutions is the fact that physical observables, such as transition probabilities and spectroscopic quadrupole moments, are calculated in the full configuration space and there is no need for effective charges. Using the bare value of the proton charge in the electric quadrupole operator, the transition probabilities between eigenvectors of the collective Hamiltonian can be directly compared with data.

In an equivalent approach the RHB binding energy surface can be mapped onto the IBM Hamiltonian. Starting from the energy surface $E_{\text{RMF}}(\beta, \gamma)$ calculated with the DD-PC1 plus separable-pairing functional, each point on the (β, γ) plane is mapped onto the corresponding point on the energy surface calculated in the IBM, referred to hereafter as $E_{\text{IBM}}(\beta_B, \gamma_B)$, using the method proposed in Ref. [2]. Here β_B and γ_B denote the boson images of the quadrupole deformation parameters β and γ, respectively, that are used as constraints in the self-consistent RHB calculation and appear as variables in the collective Hamiltonian. The boson images β_B and γ_B are related to β and γ through the proportionality $\beta_B \propto \beta$, and the equality $\gamma_B = \gamma$, respectively [1, 2]. This mapping procedure is used to determine the strength parameters of the IBM Hamiltonian.

Turning now to the IBM description, we consider the most basic type of the Hamiltonian of Eq. (2.41). The bosonic energy surface $E_{\text{IBM}}(\beta, \gamma)$ corresponds to the classical limit of the Hamiltonian, denoted by

$$E_{\text{IBM}}(\beta_B, \gamma_B) = \langle \Phi(\beta_B, \gamma_B) | \hat{H}_{\text{IBM}} | \Phi(\beta_B, \gamma_B) \rangle \qquad (5.5)$$

where $|\Phi(\beta_B, \gamma_B)\rangle$ denotes the boson coherent state of Eq. (2.36). As in our previous studies [1, 2], it is assumed that the deformation parameters for proton and neutron bosons can take identical values: $\beta_\pi = \beta_\nu \equiv \beta_B$ and $\gamma_\pi = \gamma_\nu \equiv \gamma_B$. The analytical form of $E_{\text{IBM}}(\beta_B, \gamma_B)$ can be found in Eq. (2.46). Hereafter we denote the bosonic energy surface as $E_{\text{IBM}}(\beta, \gamma)$, omitting the indices of β_B and γ_B.

The boson Hamiltonian $\hat{H}_{\text{IBM}}$, parametrized by the microscopically calculated coupling constants, is diagonalized in the $M = 0$ boson space. Here M denotes the z-component of the total boson angular momentum L. Reduced quadrupole transition probabilities $B(E2)$ are calculated for transitions between the eigenstates of the IBM Hamiltonian.

Here we point out again that the total boson energy $E_{\text{IBM}}(\beta, \gamma)$ has been related to the microscopic EDF energy surface (total energy). However, for the IBM Hamiltonian $\hat{H}_B$ one cannot make a distinction between the kinetic and potential terms, as in the corresponding collective Hamiltonian $\hat{H}_{\text{coll}}$. Nevertheless, the effects relevant to both vibrational and rotational kinetic energies are assumed to be incorporated into the IBM approach by adjusting $E_{\text{IBM}}(\beta, \gamma)$ to be as close as possible to the microscopic surface $E_{\text{RHB}}(\beta, \gamma)$. This prescription turned out to be valid for vibrational and γ-soft nuclei at moderate quadrupole deformation [1, 2], similarly to the conventional mapping method of Ref. [22]. For rotational nuclei with large quadrupole deformation, however, the overall scale of the IBM rotational spectra differs from the experimental one [1, 2]. The discrepancy partially arises because nuclear rotational properties, characterized by the overlap of the intrinsic state and the rotated one, differs from the rotational characteristics of the corresponding boson system [3]. This problem may be cured by the recently proposed prescription [3], in which the rotational response (i.e., cranking) of boson system is related to the rotational response of nucleon system. This procedure goes beyond the simple analysis of the zero-frequency energy surface. In order that the boson rotational response becomes equal to the fermion (nucleon) response, an additional kinetic term $\hat{L} \cdot \hat{L}$ has to be included in the boson Hamiltonian, with a coupling constant determined microscopically [3]. The term $\hat{L} \cdot \hat{L}$ directly influences the moment of inertia of rotational band with the eigenvalue $L(L + 1)$. However, the above-mentioned problem, concerning the IBM rotational spectra, does not occur in the considered Pt nuclei, and thus one does not need to include the $\hat{L} \cdot \hat{L}$ term in the present case.

Similar problems with the overall scale of the rotational spectra are also encountered in the collective Hamiltonian model, when the IB formula is used to calculate the moments of inertia [6–8]. The inclusion of an additional scale parameter is often necessary because of the well known fact that the IB formula predicts effective moments of inertia that are considerably smaller than empirical values. More realistic values are only obtained if one uses the Thouless-Valatin (TV) formula [23],

5.2 Bohr Hamiltonian

but this procedure is computationally much more demanding. In the present case we have used the IB moments of inertia in the calculation of excitation spectra of Pt nuclei, and the agreement with experiment is such that no renormalization of the effective moments of inertia is required. This result allows for a direct comparison of the IBM spectra to the solutions of the collective Hamiltonian.

Most deformed nuclei display axially-symmetric prolate ground-state shapes, but few areas of the nuclide chart are characterized by the occurrence of non-axial shapes. One example is the $A \approx 190$ mass region, where both prolate to oblate shape transitions, and even triaxial ground-state shapes have been predicted.

The left-hand side of Fig. 5.1 shows the self-consistent RHB quadrupole binding energy maps of the ^{192,194,196}Pt isotopes in the $\beta - \gamma$ plane, calculated with the DD-PC1 energy density functional. The energy surfaces are γ-soft, with shallow minima at $\gamma \approx 30°$. In general the equilibrium deformation decreases with mass number and, proceeding to even heavier isotopes, one finds that the energy map of ^{198}Pt has also a non-axial minimum, whereas ^{200}Pt displays a slightly oblate minimum [10], signaling the shell-closure at the neutron number $N = 126$. On the right-hand side of Fig. 5.1, we plot the corresponding IBM energy surfaces $E_{\text{IBM}}(\beta, \gamma)$, obtained by mapping each point of surface $E_{\text{RHB}}(\beta, \gamma)$ onto the energy surface calculated in the IBM, following the procedure of Ref. [2]. To be able to compare the low-energy spectra in the two models, the IBM surfaces are mapped in such a way to reproduce the RHB energy surfaces up to ≈ 2 MeV above the mean-field minimum. This means that the maps shown in Fig. 5.1 can only be compared for values of β not very different from the minimum $\beta_{\min}$. For larger values of β, that is, for higher excitation energies the topology of the RHB surfaces is determined by single-nucleon configurations that are not included in the model space (valence space) from which the IBM bosons are constructed. For large β-deformations, therefore, one should not try to map the microscopic energy surfaces onto the IBM. This is the reason why the IBM energy surfaces are by construction always rather flat in the region $\beta \gg \beta_{\min}$. In the vicinity of the minima the curvatures of the IBM energy maps are rather similar to those of the original RHB surfaces both in β and γ directions. The derived values for the χ_π and χ_ν parameters in Eq. (2.41) satisfy $\chi_\pi + \chi_\nu \sim 0$, characteristic for a γ-soft energy surface.

One might notice that the IBM energy maps reproduce the value of β at the minima predicted by the RHB calculation, whereas the mapping does not reproduce the shallow triaxial minima of the RHB surfaces. The minima of the IBM maps are either oblate or prolate. This is because the IBM Hamiltonian of Eq. (2.41) is too restricted to produce a triaxial minimum. In the analytical expression for $E_{\text{IBM}}(\beta, \gamma)$ the γ-dependent term is proportional to $(\chi_\pi + \chi_\nu) \cos 3\gamma$, and this places the minimum either on the prolate or oblate side according to the sign of $(\chi_\pi + \chi_\nu)$. The Pt nuclei considered here do not display any rapid structural change but remain γ-soft. This feature appears to be independent of the choice of the EDF. A recent microscopic calculation using the Gogny-D1S EDF [24] also yielded shallow triaxial shapes, rather flat in the oblate region [25], but quantitatively consistent with the present analysis. A similar trend was reported in other EDF-based studies of ground-state shapes of Pt isotopes [11, 26, 27]. In the present calculation the RHB surfaces become softer in γ

136 5 Comparison with Geometrical Model

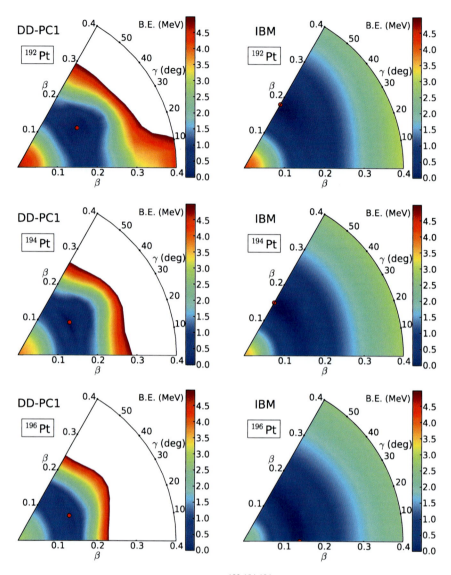

Fig. 5.1 Self-consistent binding-energy maps of ^{192,194,196}Pt in the $\beta - \gamma$ plane ($0° \leq \gamma \leq 60°$), calculated with the RHB model using the DD-PC1 functional (*left panels*), and the corresponding mapped energy surface of the IBM, $E_{\text{IBM}}(\beta_B, \gamma_B)$. The IBM total energies are depicted in terms of β and γ, where $\beta \propto \beta_B$ and $\gamma = \gamma_B$ (see text for definition). The figures have been taken from Ref. [17]

5.2 Bohr Hamiltonian

with increasing neutron number, and the softest nucleus is ^{196}Pt. The corresponding IBM energy surfaces follow this evolution, but do not reproduce the triaxial minima because of the reasons explained above. The recent Gogny-EDF calculation [25] predicts ^{192}Pt to be the softest Pt isotope in this mass region.

5.3 Geometrical and Bosonic Spectra

In Fig. 5.2 we display the corresponding low-energy collective spectra of ^{192,194,196}Pt obtained from the collective Hamiltonian (middle panels), and the IBM Hamiltonian (panels on the right). The calculated ground-state and (quasi) γ-vibration bands are compared to the corresponding sequences of experimental states [31]. The eigenstates of the collective Hamiltonian in Eq. (5.1) are completely determined by the DD-PC1 energy density functional plus a separable pairing interaction, and the transition probabilities are calculated in the full configuration space using the bare value of the proton charge. Since $\hat{H}_{\text{IBM}}$ in Eq. (2.41) acts only in the boson valence space, to calculate the $B(E2)$ values one needs two additional parameters: the proton-boson and neutron-boson effective charges. For simplicity, here we take these effective charges to be equal, and in each nucleus normalize the $B(E2)$ values obtained in the IBM to reproduce the transition probability $B(E2; 2_1^+ \to 0_1^+)$ calculated with the collective Hamiltonian. Thus we can only compare the ratios of the IBM $B(E2)$ values, divided by $B(E2; 2_1^+ \to 0_1^+)$, to those predicted by the collective Hamiltonian based on DD-PC1, and to available data.

For the ground-state band, both the collective Hamiltonian and the IBM predict excitation spectra in close agreement with experiment. For ^{192}Pt, in particular, the calculated ground-state bands seem to indicate a somewhat larger deformation than observed experimentally. In fact, the theoretical energy ratio $R_{4/2} = E(4_1^+)/E(2_1^+)$ is 2.59 with collective Hamiltonian, and is 2.69 with the IBM Hamiltonian, compared to the experimental value $R_{4/2} = 2.48$. A similar trend is also found for the other two nuclei. A more pronounced difference between the predictions of the two models is found in the E2 decay pattern of the ground-state band, particularly in ^{194}Pt nucleus for which data are available up to angular momentum 10^+. For the spectrum calculated with the collective model, the E2 transition rates from the state with angular momentum L ($L \geq 2$) to the one with $L - 2$ keep increasing as function of L, even though the corresponding experimental $B(E2)$ values in ^{192,194}Pt decrease starting from $L = 6$. The trend of the $B(E2)$ values calculated with the IBM, on the other hand, is much closer to experiment. The $B(E2)$'s decrease in the IBM because the model space is built from valence nucleons only, and the wave functions of higher angular-momentum states correspond to simple configurations of fully aligned d-bosons [12, 13], whereas there is no limit on the angular momentum of eigenstates of the collective Hamiltonian.

A more significant difference between the spectroscopic properties predicted by the collective Hamiltonian and the IBM is found in the sequence of levels built on the state 2_2^+—the (quasi) γ-band. The IBM spectra display a staggering of excitation

138   5 Comparison with Geometrical Model

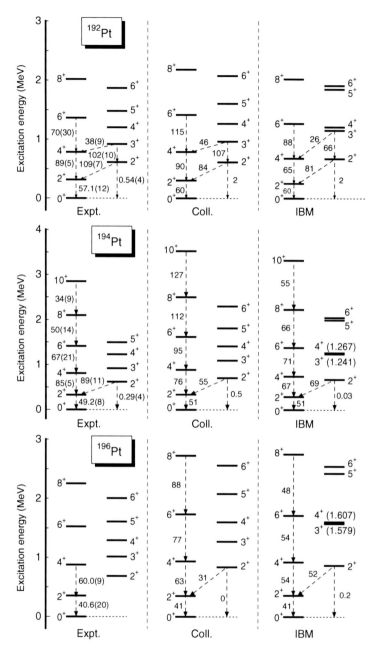

Fig. 5.2 Low-lying collective spectra of ^{192,194,196}Pt nuclei, calculated with the collective Hamiltonian based on the DD-PC1 functional and the corresponding IBM Hamiltonian, in comparison with available data. For each nucleus, the $B(E2)$ values (in Weisskopf units) obtained in the IBM are normalized to the $B(E2; 2_1^+ \to 0_1^+)$ predicted by the collective Hamiltonian. The experimental excitation spectra and $B(E2)$ values are from Refs. [31] and [32–34], respectively. The figure has been taken from Ref. [17]

energies above 2_γ^+, with the formation of doublets $(3_\gamma^+\, 4_\gamma^+), (5_\gamma^+\, 6_\gamma^+), \ldots$ etc, whereas the collective Hamiltonian yields a regular excitation pattern consistent with the experimental band. To be more precise, the IBM spectra correspond to γ-unstable nuclei, and are close to the limit of O(6) dynamical symmetry in which eigenstates of a boson Hamiltonian with the same τ quantum number are degenerate [14]. On the other hand, the γ-bands predicted by the collective model, as well as the experimental sequence, seem to be closer to rigid triaxiality [28]. The difference between the collective Hamiltonian and the IBM arises probably because the shallow triaxial minima of the RHB energy surfaces are not reproduced by the mapping onto the IBM total energy (cf. Fig. 5.1). The agreement of the IBM (quasi) γ-band with experiment could be improved by introducing additional interaction terms in the IBM Hamiltonian, i.e., three-body terms (the so-called cubic terms) [29, 30]. Terms of this type should be included for a more precise analysis and comparison of states above the yrast with experimental results. The analyses along this line will be described in detail in Chap. 6.

A nice feature of the present calculation, particularly the one with the IBM Hamiltonian, is that the predicted $B(E2)$ values for the transition $2_2^+ \to 2_1^+$ are comparable to or even larger than those corresponding to $4_1^+ \to 2_1^+$. This result is consistent with the experimental trend, whereas in the recent Gogny-based EDF calculation of Ref. [25], the $2_2^+ \to 2_1^+$ transitions were much weaker than $4_1^+ \to 2_1^+$. The corresponding Gogny energy surfaces displayed pronounced oblate minima in Ref. [25], unlike the present deformation energy maps shown in Fig. 5.1.

Finally, in Figs. 5.3 and 5.4 we compare the absolute squares of the collective wave functions for the yrast states $0_1^+, 2_1^+, 4_1^+$, and the band-head of the γ-band of ^{192}Pt, calculated in the two models. These quantities are proportional to the probability density distributions in the $\beta - \gamma$ plane. Figure 5.3 shows the distribution, denoted by $f_L(\beta, \gamma)$, for ^{192}Pt nucleus. $f_L(\beta, \gamma)$ is written as

$$f_L(\beta, \gamma) = \sum_{M=-L}^{L} |\langle \Psi_M^L | \Phi(\beta, \gamma) \rangle|^2, \tag{5.6}$$

where $|\Psi_M^L\rangle$ denotes the IBM eigenstate for the state with angular momentum L and projection M. The way to calculate the absolute square $f_L(\beta, \gamma)$ is described in Appendix B.2.

In Fig. 5.3 the wave functions of the yrast states are concentrated along the oblate axis, only for the state 4_1^+ the maximum of the absolute square is located at $\gamma \sim 55°$, and somewhat larger deviations from pure oblate configurations are found for higher angular momenta. For the state 2_2^+, on the other hand, the peak appears in the triaxial region ($\gamma \sim 35°$), and the distribution is extended more toward oblate quadrupole deformations. The rather large overlap of the collective wave functions for the states 2_1^+ and 2_2^+ explains the particularly strong $2_2^+ \to 2_1^+$ transitions in this nucleus, and similarly in the other two Pt isotopes considered here. The corresponding absolute squares of the eigenstates of the collective Hamiltonian are shown in Fig. 5.4. In this case already the wave functions of the yrast states reflect the γ-softness of the

140 5 Comparison with Geometrical Model

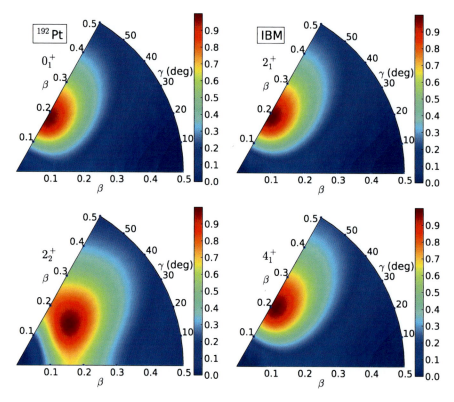

Fig. 5.3 Absolute squares of the IBM wave functions in the $\beta - \gamma$ plane for the yrast states 0_1^+, 2_1^+, 4_1^+, and the band-head of the γ-band 2_2^+ of ^{192}Pt. The positions of the maxima are denoted by a dot. The figures have been taken from Ref. [17]

RHB energy surface, and the maxima of the absolute squares are found in the triaxial region of the $\beta - \gamma$ plane.

5.4 Brief Summary

To summarize this chapter, the two models have been compared here in a study of the evolution of non-axial shapes in Pt isotopes. Starting from the binding energy surfaces of ^{192,194,196}Pt, calculated with the DD-PC1 energy density functional plus a separable pairing interaction, we have analyzed the resulting low-energy collective spectra obtained from the collective Hamiltonian, and the corresponding IBM-2 Hamiltonian. The calculated ground-state and γ-vibration bands have been also compared to the corresponding sequences of experimental states. Both models predict that excitation energies and $B(E2)$ values are in agreement with data. In particular, we notice the excellent result for the predicted excitation energy of the band-head of

5.4 Brief Summary

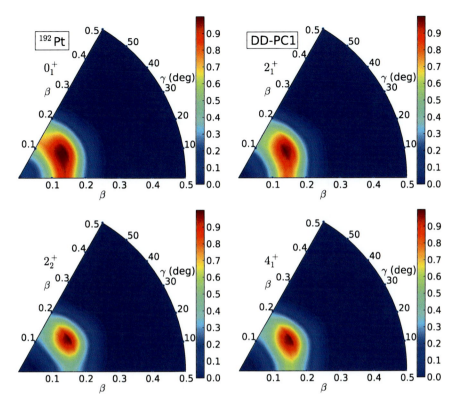

Fig. 5.4 Same as described in the caption to Fig. 5.3 but for the eigenstates of the collective Hamiltonian. The figures have been taken from Ref. [17]

the γ-band, as well as the good agreement with the experimental $B(E2)$ values for transitions between the γ-band and the yrast band.

There are also significant differences in the predictions of the two models. With the present form of the IBM Hamiltonian, restricted to two-body boson interactions, its expectation value in the boson coherent state does not reproduce the shallow triaxial minima of the binding energy maps predicted by the constrained self-consistent mean-field calculation using DD-PC1. Since the mapped IBM energy surface is γ-soft rather than triaxial, the resulting spectra display a staggering of excitation energies above 2_γ^+, with the formation of doublets $(3_\gamma^+\ 4_\gamma^+)$, $(5_\gamma^+\ 6_\gamma^+)$, ... etc, in contrast to the regular excitation pattern observed in experiment and reproduced by the collective Hamiltonian. This problem could be solved by including three-body boson terms in the IBM Hamiltonian. When considering the calculated $B(E2)$ values for transitions in the ground-state band, the IBM reproduces the gradual decrease of transition rates with angular momentum for $L \geq 6$, reflecting the finiteness of the valence space. On the other hand, even though the collective Hamiltonian predicts parameter-free $B(E2)$ values in excellent agreement with experiment for transitions between low-spin states, the calculated transition probabilities keep increasing with angular momentum, in contrast to data.

Both models are based on binding energy surfaces calculated at zero rotational frequency. In general this leads to effective rotational moments of inertia that are lower than empirical values, that is, the calculated rotational bands are stretched in energy compared to experimental bands. In the collective Hamiltonian the moments of inertia can be improved by including the Thouless-Valatin dynamical rearrangement contributions. For the IBM Hamiltonian one needs to include the kinetic rotational term [3], and perform the mapping of microscopic energy surfaces calculated at finite values of the rotational frequency. We have already started with the implementation of these modifications in our current version of the collective Hamiltonian based on relativistic EDF, and in the IBM Hamiltonian. The comparison of the improved models will be the subject of a future study.

References

1. Nomura K, Shimizu N, Otsuka T (2008) Mean-field derivation of the interacting boson model Hamiltonian and exotic nuclei. Phys Rev Lett 101:142501
2. Nomura K, Shimizu N, Otsuka T (2010) Formulating the interacting boson model by mean-field methods. Phys Rev C 81:044307
3. Nomura K, Otsuka T, Shimizu N, Guo L (2011) Microscopic formulation of the interacting boson model for rotational nuclei. Phys Rev C 83:041302(R)
4. Bonche P, Dobaczewski J, Flocard H, Heenen P-H, Meyer J (1990) Analysis of the generator coordinate method in a study of shape isomerism in ^{194}Hg. Nucl Phys A510:466
5. Delaroche J-P, Girod M, Libert L, Goutte H, Hilaire S, Peru S, Pillet N, Bertsch GF (2010) Structure of even-even nuclei using a mapped collective Hamiltonian and the D1S Gogny interaction. Phys Rev C 81:014303
6. Nikšić T, Li ZP, Vretenar D, Próchniak L, Meng J, Lalazissis GA, Ring P (2009) Beyond the relativistic mean-field approximation. III. Collective Hamiltonian in five dimensions. Phys Rev C 79:034303
7. Li ZP, Nikšić T, Vretenar D, Meng J, Lalazissis GA, Ring P (2009) Microscopic analysis of nuclear quantum phase transitions in the N≈90 region. Phys Rev C 79:054301
8. Li ZP, Nikšić T, Vretenar D, Meng J (2010) Microscopic description of spherical to γ-soft shape transitions in Ba and Xe nuclei. Phys Rev C 81:034316
9. Vretenar D, Afanasjev AV, Lalazissis GA, Ring P (2005) Relativistic Hartree Bogoliubov theory: static and dynamic aspects of exotic nuclear structure. Phys Rep 409:101
10. Nikšić T, Vretenar D, Ring P (2011) Relativistic nuclear energy density functionals: mean-field and beyond. Prog Part Nucl Phys 66:519
11. Nikšić T, Ring P, Vretenar D, Tian Y, Ma Z-y (2010) 3D relativistic Hartree-Bogoliubov model with a separable pairing interaction: triaxial ground-state shapes. Phys Rev C 81:054318
12. Arima A, Iachello F (1975) Collective nuclear states as representations of a SU(6) group. Phys Rev Lett 35:1069
13. Iachello F, Arima A (1987) The interacting boson model. Cambridge University Press, Cambridge
14. Cizewski JA, Casten RF, Smith GJ, Stelts ML, Kane WR, Borner HG, Davidson WF (1978) Evidence for a new symmetry in nuclei: the structure of ^{196}Pt and the O(6) limit. Phys Rev Lett 40:167

15. Casten RF, Cizewski JA (1978) The O(6)→rotor transition in the Pt-Os nuclei. Nucl Phys A309:477
16. Bohr A, Mottelson BR (1969, 1975) Nuclear structure, vol I single-particle motion: vol II nuclear deformations. Benjamin, New York
17. Nomura K, Nikšić T, Otsuka T, Shimizu N, Vretenar D (2011) Quadrupole collective dynamics from energy density functionals: collective Hamiltonian and the interacting boson model. Phys Rev C 84:014302
18. Nikšić T, Vretenar D, Ring P (2008) Relativistic nuclear energy density functionals: adjusting parameters to binding energies. Phys Rev C 78:034318
19. Nikšić T, Ring P, Vretenar D, Tian Y, Ma ZY (2010) 3D relativistic Hartree-Bogoliubov model with a separable pairing interaction: triaxial ground-state shapes. Phys Rev C 81:054318
20. Inglis DR (1956) Nuclear moments of inertia due to nucleon motion in a rotating well. Phys Rev 103:1786
21. Belyaev ST (1961) Concerning the calculation of the nuclear moment of inertia. Nucl Phys 24:322
22. Otsuka T, Arima A, Iachello F (1978) Shell model description of interacting bosons. Nucl Phys A309:1
23. Thouless DJ, Valatin JG (1962) Time-dependent Hartree-Fock equations and rotational states of nuclei. Nucl Phys 31:211
24. Berger JF, Girod M, Gogny D (1984) Microscopic analysis of collective dynamics in low energy fission. Nucl Phys A428:23c
25. Nomura K, Otsuka T, Rodríguez-Guzmán R, Robledo LM, Sarriguren P (2011) Structural evolution in Pt isotopes with the interacting boson model Hamiltonian derived from the Gogny energy density functional. Phys Rev C 83:014309
26. Robledo LM, Rodríguez-Guzmán R, Sarriguren P (2009) Role of triaxiality in the ground-state shape of neutron-rich Yb, Hf, W, Os and Pt isotopes. J Phys G Nucl Part Phys 36:115104
27. Rodriguez-Guzman R, Sarriguren P, Robledo LM, Garcis-Ramos JE (2010) Mean field study of structural changes in Pt isotopes with the Gogny interaction. Phys Rev C 81:024310
28. Davydov AS, Filippov GF (1958) Rotational states in even atomic nuclei. Nucl Phys 8:237
29. Heyde K, Van Isacker P, Waroquier M, Moreau J (1984) Triaxial shapes in the interacting boson model. Phys Rev C29:1420
30. Casten RF, von Brentano P, Heyde K, Van Isacker P, Jolie J (1985) The Interplay PF γ-softness and triaxiality in O(6)-like nuclei. Nucl Phys A439:289
31. Brookhaven National Nuclear Data Center (NNDC). http://www.nndc.bnl.gov/nudat2/index.jsp
32. Baglin CM (1998) Nuclear data sheets for A = 192. Nucl Data Sheets 84:717
33. Singh B (2006) Nuclear data sheets for A = 194. Nucl Data Sheets 107:1531
34. Xiaolong H (2007) Nuclear data sheets for A = 196. Nucl Data Sheets 108:1093

Chapter 6
Is Axially Asymmetric Nucleus γ Rigid or Unstable?

6.1 Overview

Nuclear shapes reflect deformations of the nuclear surface that arise from collective motion of many nucleons [1]. Ground states of most non-spherical nuclei are characterized by axially-symmetric quadrupole deformations—prolate or oblate. There are, however, also many nuclei in which axial symmetry, i.e., the invariance under the rotation around the symmetry axis of the intrinsic state, is broken in the ground state. The description of axially asymmetric shapes and the resulting triaxial quantum many-body rotors is not restricted to nuclear physics, but has also been developed for other finite quantum systems like polyatomic molecules [2], and hence presents a topic of broad interest.

To analyze the variation of ground-state shapes in a sequence of nuclei as, for instance, in an isotopic chain that extends to exotic short-lived isotopes far from stability, it is essential to provide a quantitative microscopic description of deformations characterized by both axial and triaxial mass quadrupole moments. The quadrupole moments can be related to the polar deformation parameters β and γ. The parameter β is proportional to the intrinsic quadrupole moment, and the angular variable γ specifies the type and orientation of the shape. The limit $\gamma = 0$ corresponds to axial prolate shapes, whereas the shape is oblate for $\gamma = \pi/3$. Triaxial shapes are associated with intermediate values $0 < \gamma < \pi/3$. Such shapes have been investigated extensively using theoretical approaches that are essentially based on the rigid-triaxial rotor model of Davydov and Filippov [3] and the γ-unstable rotor model of Wilets and Jean [4]. The former assumes that the collective potential has a stable minimum for a particular value of γ [1], whereas in the latter the potential does not depend on γ and thus the collective wave functions are spread out in the γ direction.

However, presumably all known axially-asymmetric nuclei exhibit features that are almost exactly in between these two geometrical limits, characterized by the energy-level pattern of quasi-γ band: relative locations of the odd-spin to the even-spin levels. As the two models originate from different physical pictures, the question of whether axially-asymmetric nuclei are γ rigid or unstable has attracted considerable theoretical interest [1, 5–8]. In this chapter we address this question from a microscopic perspective, and identifies the appropriate Hamiltonian of the interacting boson model (IBM) [6, 7] for γ-soft nuclei, consistent with the microscopic picture. We thereby provide a solution to the problem concerning the energy-level pattern of the odd-spin states.

The most complete and accurate microscopic description of ground-state properties and collective excitations over the whole nuclide chart is presently provided by the framework of Energy Density Functionals (EDFs). Both non-relativistic [9–11], and relativistic [12, 13] EDFs have successfully been employed in numerous studies on the rich variety of shape phenomena, and the resulting complex excitation spectra and decay patterns across the entire chart of nuclides [14–17]. The starting point is usually a constrained self-consistent mean-field calculation of the binding-energy surface with the mass quadrupole moments as constrained quantities [5]. This is illustrated in the first row of Fig. 6.1, where we display the self-consistent quadrupole binding-energy maps of ^{134}Ba (a) and ^{190}Os (b) in the β–γ plane. The binding energy surface of ^{134}Ba is calculated using the relativistic Hartree-Bogoliubov model [12] with the DD-PC1 [18] functional, and that of ^{190}Os employing the Hartree-Fock plus BCS model [19] with the Skyrme functional SkM* [20]. These functionals are representative of the two classes—relativistic and non-relativistic EDFs, and will be used throughout this work to demonstrate that the principal conclusions do not depend on the particular choice of the EDF. One notices that in both cases the potential is very soft in the γ degree of freedom, with ^{134}Ba displaying a nearly γ-independent picture, whereas a more pronounced rigid triaxial shape is predicted for ^{190}Os with the minimum at $\gamma \approx 30°$.

To calculate excitation spectra and transition rates, it is necessary to project from the mean-field solution states with good quantum numbers and take into account fluctuations around the mean-field minimum. Symmetry restoration and fluctuations of quadrupole deformation can be treated simultaneously by mixing projected states that correspond to different intrinsic configurations. An effective approach for configuration mixing calculations is the generator coordinate method (GCM) [5], with multipole moments used as coordinates that generate the intrinsic wave functions. GCM configuration mixing of axially symmetric states has routinely been employed in structure studies, but the application of this method to triaxial shapes presents a much more involved and technically difficult problem.

From the viewpoint of the interacting boson model [6, 7], it is well know that the O(6) dynamical symmetry [22] embodies systems with γ-independent collective potentials. The geometrical picture of the O(6) limit of the IBM emerges using the coherent-state framework [23]. The coherent state represents the intrinsic wave function of the boson system, and O(6) states in the laboratory system can be generated by angular momentum projection [23]. The triaxial-rotor features of the IBM were emphasized already in [24, 25], leading to the "equivalence" ansatz of the γ-rigid and the O(6) descriptions of the low-lying spectra [26].

6.2 Three-Body Boson Term

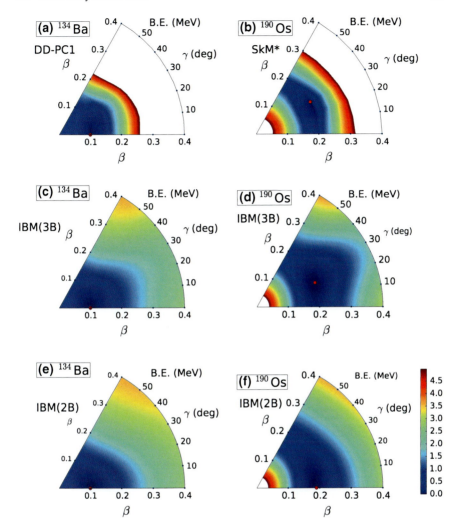

Fig. 6.1 Self-consistent binding-energy maps of ^{134}Ba (**a**) ^{190}Os (**b**), calculated using the relativistic Hartree-Bogoliubov model with the DD-PC1 [18] functional, and the Hartree-Fock plus BCS model with the Skyrme functional SkM* [20], respectively. The corresponding mapped energy surfaces of the IBM are plotted in the *middle row* (for the full IBM Hamiltonian Eq. (6.1) that contains the three-body term), and in the *lower row* (for the IBM Hamiltonian without the three body term). The figure is taken from Ref. [21]

6.2 Three-Body Boson Term

This study is performed using the proton-neutron IBM (IBM-2), which includes proton (neutron) monopole s_π (s_ν) and quadrupole d_π (d_ν) bosons, representing an effective collective approximation of $J = 0^+$ and 2^+ valence proton (neutron) pairs, respectively [27]. The number N_π (N_ν) of proton (neutron) bosons equals the number of valence proton (neutron) pairs (particles or holes), with respect to the nearest proton (neutron) closed shell [27]. The following IBM-2 Hamiltonian is employed:

$$\hat{H}_{\text{IBM}} = \varepsilon(\hat{n}_{d\pi} + \hat{n}_{d\nu}) + \kappa \hat{Q}_\pi \cdot \hat{Q}_\nu + \hat{H}_{3\text{B}} \tag{6.1}$$

with the d-boson number operator $\hat{n}_{d\rho} = d_\rho^\dagger \cdot \tilde{d}_\rho$, and the quadrupole operator $\hat{Q}_\rho = s_\rho^\dagger \tilde{d}_\rho + d_\rho^\dagger s_\rho + \chi_\rho [d_\rho^\dagger \tilde{d}_\rho]^{(2)}$, ($\rho = \pi, \nu$). The third term $\hat{H}_{3\text{B}}$ represents a three-body boson interaction:

$$\hat{H}_{3\text{B}} = \sum_{\rho \neq \rho'} \sum_L \theta_L^\rho [d_\rho^\dagger d_\rho^\dagger d_{\rho'}^\dagger]^{(L)} \cdot [\tilde{d}_{\rho'} \tilde{d}_\rho \tilde{d}_\rho]^{(L)}. \tag{6.2}$$

Three-body terms of this type have previously been employed in the IBM phenomenology [28], but here for the first time it is used in the microscopic IBM-2 framework. Three-body terms could in general have other combinations of proton and neutron d-boson operators. However, since the proton-neutron quadrupole interaction dominates over the proton-proton and neutron-neutron ones for medium-heavy and heavy nuclei, the present form of Eq. (6.2) provides a very good approximation for the three-body boson interaction. For each ρ and ρ', there are five linearly independent combinations in Eq. (6.2), determined by the value of $L = 0, 2, 3, 4, 6$ [29]. However, only the term with $L = 3$ gives rise to a stable triaxial minimum at $\gamma \approx 30°$ [28], because its expectation value in the classical limit is proportional to $\cos^2 3\gamma$. We thus consider only the $L = 3$ in Eq. (6.2) and, in addition, assume $\theta_3^\pi = \theta_3^\nu \equiv \theta_3$. The form of the three-body boson term is discussed in more detail in Appendix B.1.

The parameters $\varepsilon, \kappa, \chi_\pi, \chi_\nu$ and θ_3 are fixed, following the procedure of Ref. [17]: the microscopic quadrupole binding energy surface, obtained from a mean-field calculation using a given EDF, is mapped onto the corresponding boson energy surface, i.e., expectation value of $\hat{H}_{\text{IBM}}$ in the coherent state (cf. [17, 30] for details). The deduced value of $\theta_3 > 0$ varies gradually with boson number: $|\theta_3/\kappa| \approx 1$ for $1 \leq N_\pi + N_\nu \leq 5$ and ≈ 0.5 for $5 \leq N_\pi + N_\nu \leq 10$.

The two-body IBM Hamiltonian cannot produce deformed binding energy surfaces with stable triaxial minima. Its coherent-state expectation value either has a minimum at $\gamma = 0°$ (prolate shapes) or $60°$ (oblate shapes), or is independent of γ in the O(6)-limit. This is nicely illustrated in Fig. 6.1 where the mapped energy surfaces of the IBM are plotted in the middle row (for the full IBM Hamiltonian Eq. (6.1) that contains the three-body term), and in the lower row (for the IBM Hamiltonian without the three body term). For the ^{190}Os the Hartree-Fock plus BCS model with the Skyrme functional SkM* predicts minimum at $\gamma \approx 30°$, that can only be reproduced on the mapped surface that correspond to the expectation valued of the full IBM Hamiltonian with the three-body term (panel (d) in Fig. 6.1). The two-body

6.2 Three-Body Boson Term

IBM Hamiltonian yields a binding energy surface that is soft in the γ-direction, but the minimum is on the $\gamma = 0°$ axis (panel (f)).

A distinction between γ-soft and rigid triaxial nuclei arises when considering the ratio of excitation energies [8]:

$$S(J, J-1, J-2) \equiv \frac{[\{E(J) - E(J-1)\} - \{E(J-1) - E(J-2)\}]}{E(2_1^+)} \quad (6.3)$$

for the γ ($K^\pi = 2^+$) band $J^\pi = 2_\gamma^+, 3_\gamma^+, 4_\gamma^+ \ldots$ The excitation energies $E(J)$ are obtained by diagonalization of the boson Hamiltonian Eq. (6.1), and the quadrupole operators Q_ρ are used in the calculation of E2 transition rates, with identical proton and neutron boson effective charges.

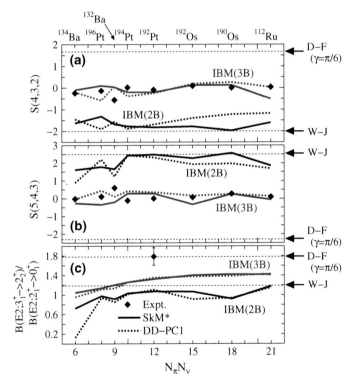

Fig. 6.2 The energy ratios $S(4, 3, 2)$ (**a**) and $S(5, 4, 3)$ (**b**), and $B(E2; 3_1^+ \to 2_2^+)/B(E2; 2_1^+ \to 0_1^+)$ (**c**), as functions of the product of the number of proton and neutron bosons $N_\pi N_\nu$, for a set of typically non-axial medium-heavy and heavy nuclei. In each panel IBM(3B) and IBM(2B) denote results obtained with the full IBM Hamiltonian including the three-body term, and with only two-body boson terms. Skyrme SkM* and relativistic DD-PC1 functionals are used. Results are compared to available data [31–47], and to the predictions of the rigid-triaxial rotor model (D-F) and the γ-unstable rotor model (W-J)

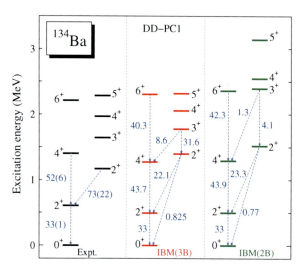

Fig. 6.3 Low-lying spectra and $B(E2)$ values (in Weisskopf units) of ^{134}Ba. The bands calculated with IBM Hamiltonians with and without the three-body term Eq. (6.2), are compared to experimental results [48]. DD-PC1 relativistic energy density functional is used, and the boson effective charge is adjusted to reproduce the experimental value $B(E2; 2_1^+ \to 0_1^+)$

For a set of typically non-axial medium-heavy and heavy nuclei, in Fig. 6.2 we plot the energy ratios $S(4, 3, 2)$ (a) and $S(5, 4, 3)$ (b), as functions of the product of the number of proton and neutron bosons $N_\pi N_\nu$. The latter quantity reflects the amount of valence proton-neutron correlation, and hence the increase of $N_\pi N_\nu$ corresponds to an enhancement of collectivity [8]. In this chapter we consider non-axial nuclei in the mass regions $A \sim 110$, 130 and 190, whose spectra display signatures of γ-softness. The set of nuclei shown in Fig. 6.2 has been selected so that the corresponding values of $N_\pi N_\nu$ evenly span the widest possible range. The IBM excitation spectra have been calculated starting from self-consistent mean-field binding energy maps that correspond to the two functionals, Skyrme SkM* and the relativistic DD-PC1. The two energy ratios, calculated with and without the three-body term Eq. (6.2) in the IBM Hamiltonian, are plotted in comparison to data and the predictions of the rigid-triaxial rotor model of Davydov and Filippov [3] and the γ-unstable rotor model of Wilets and Jean [4]. One notices that for all considered nuclei data can only be reproduced with the IBM Hamiltonian that includes the three-body term Eq. (6.2). Both the empirical and calculated ratios fall almost exactly in between the limits of the γ-unstable rotor and the rigid-triaxial rotor models: the Wilets-Jean limit is -2.00 and the Davydov-Filippov limit is 1.67 for $S(4, 3, 2)$; the Wilets-Jean model predicts 2.50, and Davydov-Filippov -2.30 for $S(5, 4, 3)$. The IBM Hamiltonian that contains only two-body terms cannot reproduce the empirical values and, in both cases, yields energy ratios that are close to the predictions of the γ-unstable rotor model.

While the energy ratios are largely independent of the product of boson numbers, $B(E2)$ systematics reflects the evolution of collectivity. For instance, the ratio $B(E2; 3_1^+ \to 2_2^+)/B(E2; 2_1^+ \to 0_1^+)$ plotted in the lower panel of Fig. 6.2, gradually increases with $N_\pi N_\nu$. For nuclei with typically low $N_\pi N_\nu$ (≤ 10), like ^{132,134}Ba and ^{194,196}Pt, the average valued of γ is close to $0°$ or $60°$. In this case the ratio $B(E2; 3_1^+ \to 2_2^+)/B(E2; 2_1^+ \to 0_1^+)$ is closer to the Wilets-Jean limit (O(6) in the IBM representation) of 1.19. As the collectivity evolves with $N_\pi N_\nu \geq 12$, this $B(E2)$ ratio, calculated with the full IBM Hamiltonian that includes the three-body term, saturates between the γ-rigid limit of 1.78 and the γ-unstable limit of 1.19, in agreement with behavior of the energy ratios $S(J, J-1, J-2)$. The $B(E2)$ ratio calculated with only the two-body boson Hamiltonian remains close to the O(6) limit even for large values of $N_\pi N_\nu$. The feature of the $B(E2)$ pattern naturally indicates that, to measure the γ softness, one should not solely look at the excitation energies, but that the wave functions should be inspected also.

As the ratios shown in Fig. 6.2 are calculated using two completely different microscopic density functionals, it appears that the basic features of this study are not sensitive to the particular choice of the underlying EDF. It has to be emphasized the empirical values are only reproduced with the full IBM Hamiltonian, including the three-body term.

In the IBM picture, the number of proton (neutron) bosons equals half the number of the corresponding valence particle or hole pairs. Empirical systematics indicates that γ-softness mostly emerges when N_π and/or N_ν correspond to hole pairs counted from the nearest closed shells (e.g. [8]), that is, when the occupancy of major shells exceeds 50 %. All nuclei considered in this work belong to this category. Nuclei with relatively large $N_\pi N_\nu$ (≥ 12), in many of which both N_π and N_ν correspond to hole configurations, are more likely to exhibit pronounced γ-rigidity, compared to systems with low $N_\pi N_\nu$ (≤ 10). In most of the latter cases N_π and N_ν correspond to particle and hole configurations, respectively, and vice versa.

The discussion so far has focused on systematic of energy ratios and transition rates. The model, however, provides an equally accurate description of complete low-energy excitation spectra in individual nuclei. Absolute excitation energies are described precisely as well. This is highlighted by the level scheme of ^{190}Os in Fig. 6.4. Again we compare results obtained from the three-body Hamiltonians with those from two-body one and with available data [31–47]. The full IBM Hamiltonian nicely reproduces both the excitation energies and transition rates for the ground-state band and the band built on 2_2^+ (γ-band). We notice the marked effect of the three-body term on the γ-band: all states are lowered in energy but, in particular, the pronounced lowering of the odd-spin states, e.g. 3_1^+, 5_1^+ ..., breaks the quasi-degeneracy of the doublets $(3_1^+, 4_2^+)$, $(5_1^+, 6_2^+)$, etc. The lowest 3^+ level is predicted too high in many IBM calculations for γ-soft nuclei, and the empirical lowering was often ascribed to a possible coupling to two-quasiparticle states. It is evident that such mechanism might play only a minor role. These doublets (τ-multiplets) are characteristic of the γ-unstable O(6) symmetry limit of the IBM model [6, 7]. We emphasize that there are no additional adjustable parameters for the levels, that is, the parameters are completely determined by the choice of the microscopic functional

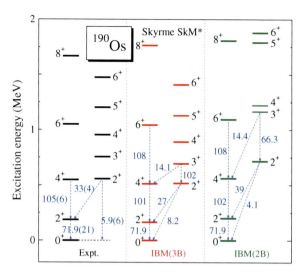

Fig. 6.4 Same as Fig. 6.3, but for ^{190}Os nucleus with Skyrme SkM* functional. Experimental data are taken from Ref. [42]. The figure is taken from Ref. [21]

and the mapping procedure. Results of similar level of agreement with experiment are also obtained in the calculation of spectra of other nuclei considered in this study.

The feature of the B(E2) pattern naturally indicates that, to measure the γ softness, one should not solely look at the excitation energies of the γ band, but that the wave functions should be inspected also. We then examine the IBM wave functions, comparing a γ-unstable system ^{134}Ba with a more rigid one ^{190}Os. Figure 6.5 shows the distributions of the wave functions of the states in the γ band, $2_2^+, 3_1^+$ and 4_2^+, in $\beta\gamma$ planes. The distribution means the absolute square of the overlap defined in Eq. (5.6), $f_L(\beta, \gamma) = \sum_{M=-J}^{J} |\langle \Phi(\beta, \gamma) | \Psi_M^J \rangle|^2$, where $|\Psi_M^J\rangle$ stands for the eigenstate of $\hat{H}_{\text{IBM}}$ (6.1) with J and $J_z = M$, and $|\Phi(\beta, \gamma)\rangle$ the intrinsic wave function. It can be shown that the wave function of a ground-band state more or less reflects the topology of the energy surface. Namely, the wave function tends to have a peak at the same location as the energy minimum. However, the γ-bandhead 2_2^+ state is spread along the oblate ($\gamma = 60°$) axis with peak at the triaxial region different from the minimum position of the energy surface. For ^{190}Os nucleus the 2_2^+ state looks rather similar in topology to the 3_1^+ one to a greater extent than for ^{134}Ba. The similarity is a possible evidence for the strong E2 transition between these states. The states in the γ band of ^{192}Os nucleus appear to have pure $K^\pi = 2^+$ nature, while the configuration mixing is much important for ^{134}Ba. In fact the 4_2^+ wave function is slightly spread out in ^{192}Os nucleus, but is more rich in topology in ^{134}Ba nucleus with a peak at $\gamma \approx 25°$.

6.3 Brief Summary

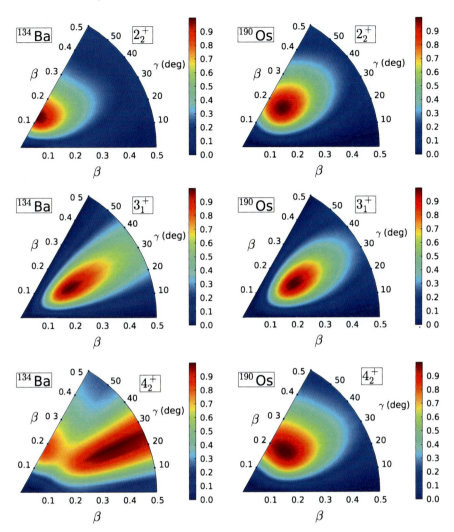

Fig. 6.5 Distributions of the absolute squares $f_L(\beta, \gamma)$ (for definition, see the main text) for the IBM wave functions in $\beta\gamma$ surfaces for the 2_2^+, 3_1^+ and 4_2^+ excited states of the nuclei ^{134}Ba with DD-PC1 (*left panels*) and ^{192}Os with SkM* ((*right panels*)

6.3 Brief Summary

To conclude, we have investigated the emergence of γ softness in atomic nuclei starting from the microscopic framework of energy density functionals. For a wide range of typically non-axial medium-heavy and heavy nuclei certain observables allow us, in comparison to microscopic calculations, to differentiate two limiting geometrical pictures: the rigid-triaxial and the γ-unstable rotors. The present analysis clearly demonstrates that neither of these pictures is realized in actual nuclei. Typical non-axial medium-heavy and heavy nuclei lie almost exactly in the middle between the two geometrical limits, as a robust regularity. In the IBM framework the regularity arises naturally only when a three-body boson interaction is included. This result points to the origin of the three-body boson interaction, suggesting the optimal IBM description of γ-soft nuclei. The principal results presented in this chapter do not depend on details nor choice of the EDF, and suggest us a comprehensive picture of triaxial shapes of atomic nuclei in a fully microscopic way, including a solution to the longstanding problem of the energy-level pattern of odd-spin states.

References

1. Bohr A, Mottelson BR (1969, 1975) Nuclear structure, vol I Single-particle motion: vol. II Nuclear deformations. Benjamin, New York
2. Herzberg G (1945) Molecular spectra and molecular structure, vol II. Van Nostrand, New York
3. Davydov AS, Filippov GF (1958) Rotational states in even atomic nuclei. Nucl Phys 8:237
4. Wilets L, Jean M (1956) Surface oscillations in even-even nuclei. Phys Rev 102:788
5. Ring P, Schuck P (1980) The nuclear many-body problem. Springer, Berlin
6. Arima A, Iachello F (1975) Collective nuclear states as representations of a SU(6) group. Phys Rev Lett 35:1069
7. Iachello F, Arima A (1987) The interacting boson model. Cambridge University Press, Cambridge
8. Casten RF (1990) Nuclear structure from a simple perspective. Oxford University Press, Oxford
9. Bender M, Heenen P-H, Reinhard P-G (2003) Self-consistent mean-field models for nuclear structure. Rev Mod Phys 75:121–180
10. Erler J, Klüpfel P, Reinhard P-G (2011) Self-consistent nuclear mean-field models: example Skyrme Hartree Fock. J Phys G Nucl Part Phys 38:033101
11. Decharge J, Girod M, Gogny D (1975) Self consistent calculations and quadrupole moments of even Sm isotopes. Phys Lett B 55:361
12. Vretenar D, Afanasjev AV, Lalazissis GA, Ring P (2005) Relativistic Hartree Bogoliubov theory: static and dynamic aspects of exotic nuclear structure. Phys Rep 409:101
13. Nikšić T, Vretenar D, Ring P (2011) Relativistic nuclear energy density functionals: mean-field and beyond. Prog Part Nucl Phys 66:519
14. Bender M, Heenen P-H (2008) Configuration mixing of angular-momentum and particle-number projected triaxial Hartree-Fock-Bogoliubov states using the Skyrme energy density functional. Phys Rev C 78:024309
15. Rodríguez TR, Egido JL (2010) Triaxial angular momentum projection and configuration mixing calculations with the Gogny force. Phys Rev C 81:064323
16. Nikšić T, Vretenar D, Lalazissis GA, Ring P (2007) Microscopic description of nuclear quantum phase transitions. Phys Rev Lett 99:092502
17. Nomura K, Shimizu N, Otsuka T (2008) Mean-field derivation of the interacting boson model Hamiltonian and exotic nuclei. Phys Rev Lett 101:142501
18. Nikšić T, Vretenar D, Ring P (2008) Relativistic nuclear energy density functionals: adjusting parameters to binding energies. Phys Rev C 78:034318
19. Bonche P, Flocard H, Heenen PH (2005) Solution of the Skyrme HF+BCS equation on a 3D mesh. Comput Phys Commun 171:49
20. Bartel J, Quentin Ph, Brack M, Guet C, Håkansson H-B (1982) Towards a better parametrisation of Skyrme-like effective forces: a critical study of the SkM force. Nucl Phys A386:79
21. Nomura K, Shimizu N, Vretenar D, Nikšić T, Otsuka T (2012) Robust regularity in γ-soft nuclei and its novel microscopic realization. Phys Rev Lett 108:132501

22. Arima A, Iachello F (1979) Interacting boson model of collective states: IV. The O(6) limit. Ann Phys 123:468–492
23. Ginocchio JN, Kirson MW (1980) Relationship between the Bohr collective Hamiltonian and the interacting-boson model. Phys Rev Lett 44:1744
24. Casten RF, Aprahamian A, Warner DD (1984) Axial asymmetry and the determination of effective γ values in the interacting boson approximation. Phys Rev C 29:356
25. Castaños O, Frank A, Van Isacker P (1984) Effective triaxial deformations in the interacting-boson model. Phys Rev Lett 52:263
26. Otsuka T, Sugita M (1987) Equivalence between γ instability and rigid triaxiality in finite boson systems. Phys Rev Lett 59:1541
27. Otsuka T, Arima A, Iachello F (1978) Shell model description of interacting bosons. Nucl Phys A309:1
28. Heyde K, Van Isacker P, Waroquier M, Moreau J (1984) Triaxial shapes in the interacting boson model. Phys. Rev. C29:1420
29. Van Isacker P, Chen J-Q (1981) Classical limit of the interacting boson Hamiltonian. Phys Rev C24:684
30. Nomura K, Shimizu N, Otsuka T (2010) Formulating the interacting boson model by mean-field methods. Phys Rev C 81:044307
31. Brookhaven National Nuclear Data Center (NNDC). http://www.nndc.bnl.gov/nudat2/index.jsp
32. Baglin CM (2002) Nucl Data Sheets 96:611
33. Balraj S (1995) Nucl Data Sheets 75:199
34. Browne E, Junde H (1999) Nucl Data Sheets 87:15
35. Basunia MS (2006) Nucl Data Sheets 107:791
36. Achterberg E et al (2009) Nucl Data Sheets 110:1473
37. Wu S-C, Niu H (2003) Nucl Data Sheets 100:483
38. Singh B, Firestone RB (1995) Nucl Data Sheets 74:383
39. Baglin CM (2010) Nucl Data Sheets 111:275
40. Baglin CM (2003) Nucl Data Sheets 99:1
41. Singh B (2002) Nucl Data Sheets 95:387
42. Singh B (2003) Nuclear Data Sheets for A = 190. Nucl Data Sheets 99:275
43. Baglin CM (1998) Nucl Data Sheets 84:717
44. Singh B (2006) Nucl Data Sheets 107:1531
45. Xiaolong H (2007) Nucl Data Sheets 108:1093
46. Xiaolong H (2007) Nucl Data Sheets 110:2533
47. Kondeva FG, Lalkovski S (2007) Nucl Data Sheets 108:1471
48. Sonzogni AA (2004) Nuclear Data Sheets for A = 134. Nucl Data Sheets 103:1

Chapter 7
Ground-State Correlation

7.1 Binding and Two-Neutron Separation Energies

We start with the Hamiltonian $\hat{H}_{\text{tot}}$ of Eq. (2.37), taking into account the global term E_0. The global term has little to do with the spectroscopy and thus has been neglected in most of the spectroscopic analysis. Here E_0 is constant for an individual nucleus and its value is determined so that the minimum of $\langle \hat{H}_{\text{tot}} \rangle$ matches the total energy of the mean-field ground state, which is denoted by E_{MF}. Namely,

$$E_0 = E_{\text{MF}} - \langle \hat{H}_{\text{IBM}} \rangle_{\min}, \qquad (7.1)$$

where $\langle \hat{H}_{\text{IBM}} \rangle_{\min}$ stands for the minimum of $\langle \hat{H}_{\text{IBM}} \rangle$ in Eq. (2.46). E_{MF} is the energy of the mean-field intrinsic state. By denoting the ground-state eigenvalue of $\hat{H}_{\text{IBM}}$ in Eq. (2.41) as E_{IBM},[1] the total energy of the IBM system, denoted by E_{tot}, is written as

$$E_{\text{tot}} = E_0 + E_{\text{IBM}}. \qquad (7.2)$$

Namely, E_{tot} is nothing but the energy eigenvalue of $\hat{H}_{\text{tot}}$ in Eq. (2.37) for the ground state in the laboratory system, which should be compared with the experiment.

Figure 7.1a, b shows the evolution of E_0, E_{tot} and experimental [1] binding energy E_{expt} as functions of the neutron number N for Sm isotopes, calculated with the Skyrme SLy4 and SkM* interactions, respectively. The boson number is taken to be $N_\nu = (N - 82)/2$ for $N > 82$, while $N_\nu = (82 - N)/2$ for $N < 82$.

It may be of interest to see N_ν-dependencies of E_0 from a simple perspective. E_0 can be approximated, for simplicity, by a polynomial

$$\bar{E}_0 = c_0 + c_1 N_\nu + c_2 N_\nu (N_\nu - 1). \qquad (7.3)$$

---

[1] The form of the IBM Hamiltonian used in this chapter takes the most simple one in Eq. (2.41) because other terms do not change the qualitative feature of the ground-state observables.

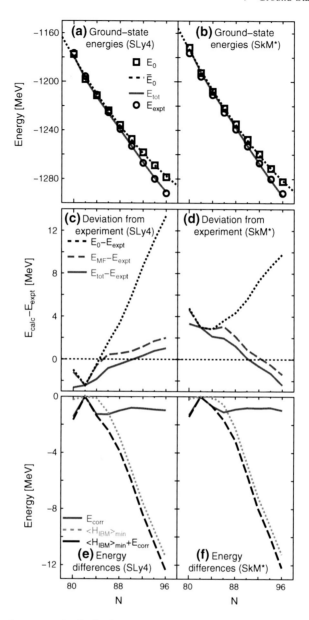

Fig. 7.1 Ground-state energies for Sm isotopes as functions of N, studied with Skyrme SLy4 (*left*) and SkM* (*right*) forces. **a**, **b** E_0, $\bar{E}_0$ (defined in Eq. (7.3)), E_{tot} and experimental [1] binding energy. **c**, **d** Deviations of the calculated ground-state energies from experiment. **e**, **f** $\langle \hat{H}_{IBM} \rangle_{min}$ (contribution from deformation), E_{corr} (correlation energy) and their sum. The figure is taken from Ref. [3]

7.1 Binding and Two-Neutron Separation Energies

Here c_0 represents the contribution from the inert core and c_1 corresponds to the energy needed to remove one neutron boson in a mean potential. The coefficients c_i's are common for all Sm isotopes in $N = 80$–96. We obtain the coefficients in MeV as $c_0 = -1195.510$ (-1190.781), $c_1 = -15.073$ (-16.393) and $c_2 = 0.476$ (0.494) for SLy4 (SkM*) case by a χ-square fit to E_0. The present value of $c_1/2$ is approximately equal to 8 MeV, which is consistent with the empirical value of one-nucleon separation energy. The fitted functions $\bar{E}_0$'s are shown also in Fig. 7.1a, b by dotted curves.

We are back now to the original definition of E_0 shown in Eq. (7.1). Figure 7.1a, b illustrates that the experimental binding energies can be reproduced to a good extent only by E_0 in the vicinity of the closed shell $N = 82$, but deviations from experiments become notable for larger N. E_{tot} reproduces the trend of the experimental values quite nicely, including those far away from the shell closure.

E_{MF} is not plotted in Fig. 7.1a, b, since it cannot be distinguished from E_{tot} in the energy scale there. We then perform more precise analyses. In Fig. 7.1c, d, the deviations of the calculated energies E_0, E_{MF} and E_{tot} from the experimental data are depicted as functions of N. The deviations of E_{MF} and E_{tot} are much smaller than that of E_0, particularly for open-shell nuclei, while the calculated results show weak dependence on the parametrizations of the Skyrme functional.

What is of particular importance in Fig. 7.1c, d is that, while E_{MF} has a kink at the closed shell $N = 82$, E_{tot} evolves smoothly as a function of N even at $N = 82$ and becomes closer to the experimental trend. For both SLy4 and SkM* cases, E_{tot} is lower in energy than E_{MF} all the way by approximately 1 MeV, and this energy difference is largest from nearly spherical to transitional regions, in which the quantum fluctuation effect seems to be most enhanced. In the following, we refer to the energy difference as the *correlation energy* E_{corr}, which is written as

$$E_{\text{corr}} \equiv E_{\text{tot}} - E_{\text{MF}}$$
$$= E_{\text{tot}} - (E_0 + \langle \hat{H}_{\text{IBM}} \rangle_{\min}). \quad (7.4)$$

Here E_{corr} is the difference between the IBM ground-state energy in the laboratory frame and the energy expectation value in the mean field, and represents quantum-mechanical contributions in the IBM. The quantity $\langle \hat{H}_{\text{IBM}} \rangle_{\min}$ can be interpreted as the deformation contribution to the ground-state energy.

We show in Fig. 7.1e, f the correlation energy E_{corr}, $\langle \hat{H}_{\text{IBM}} \rangle_{\min}$ and their sum, which corresponds to E_{IBM}, as functions of N. From around the closed shell $N = 82$ to the transitional region ($N = 88$ or 90), the correlation effect appears to be enhanced in comparison to the deformation contribution $\langle \hat{H}_{\text{IBM}} \rangle_{\min}$, while E_{corr} becomes somewhat smaller and remains constant all the way from the transitional region toward the middle of the major shell. $\langle \hat{H}_{\text{IBM}} \rangle_{\min}$ increases as a function of N, and accounts for the most part (more than 90%) of E_{IBM} for $N \geq 92$, which is obtained by the diagonalization of $\hat{H}_{\text{IBM}}$. In other words, for the strong deformation the mean-field model can give, to a certain extent, a reasonable description of the experimental binding energy.

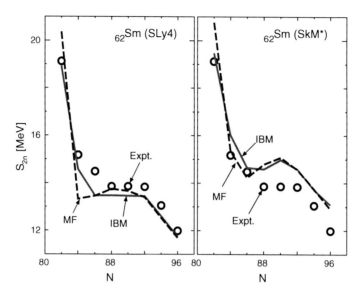

Fig. 7.2 Two-neutron separation energies for Sm isotopes for $N = 82$–96, studies with SLy4 (*left panel*) and SkM* (*right panel*) forces. Experimental data are taken from Ref. [1]. Note that the IBM results are depicted as functions of the neutron number N, instead of the number of neutron bosons N_ν. The figure is taken from Ref. [3]

The two-neutron separation energy for IBM (denoted by S_{2n}) can be calculated as a function of N_ν with N_π being fixed:

$$S_{2n}(N_\nu) = -E_{\text{tot}}(N_\nu + 1) + E_{\text{tot}}(N_\nu). \quad (7.5)$$

Note that in case the neutrons surpass a midshell, where the number of the neutron bosons are counted as that of pairs of neutron holes from the upper end of the major shell, one has to replace S_{2n} of Eq. (7.5) with its minus. Similarly, the two-neutron separation energy for the mean field (denoted by S_{2n}^{MF}) is written as a function of N with fixed Z:

$$S_{2n}^{\text{MF}}(N) = -E_{\text{MF}}(N + 2) + E_{\text{MF}}(N). \quad (7.6)$$

We show in Fig. 7.2 the two-neutron separation energies as functions of N calculated with Skyrme SLy4 and SkM* functionals. Note that the IBM separation energy S_{2n} is depicted as a function of N, not that of the neutron-boson number $N_\nu (= (N - 82)/2)$. Both Skyrme forces give similar systematics to the experiments, including the shell gap at $N = 82$ and the plateau from $N = 88$ to 92 which reflects the first-order phase transition from spherical to deformed shapes [2]. Some notable improvement by the inclusion of the correlation effect can be found around the shell closure $N = 82$. Indeed, while S_{2n}^{MF} with SLy4 deviates considerably from the experiment at $N = 84$, S_{2n} appears to be much more consistent with the experiment. This is closely associated with the finding in Fig. 7.1c, d that the kink which appears

7.1 Binding and Two-Neutron Separation Energies

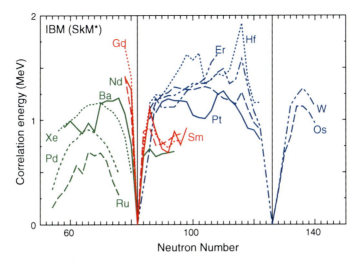

Fig. 7.3 Systematics of the correlation energies as functions of neutron number N for wide variety of medium-heavy and heavy nuclei. Skyrme SkM* functional is used. The figure is taken from Ref. [11]

around the closed shell $N = 82$ at the mean-field level is eliminated by the inclusion of the correlation energy.

Figure 7.3 shows the systematics of the correlation energies E_{corr} for a number of medium-heavy and heavy nuclei as functions of neutron number N. The correlation energy E_{corr} in this case is calculated with the IBM Hamiltonian based on the constrained HF+BCS method with Skyrme SkM* functional. A sound result is obtained similar to the Sm isotopes examined in Fig. 7.1. One observes that the correlation energy E_{corr} becomes maximal around transitional regions, e.g., in Sm isotopes with $N = 86$–90 and the Pt isotopes with $N = 90$–96, corresponding to vibrational-to-deformed shape transition, and the Pt isotopes with $N = 106$–114, corresponding to prolate-to-oblate shape transition, which is compatible with the discussion in Sect. 4.3.1.

Here we note other studies discussing systematic trend of the correlation energy for a large number of nuclei in terms of symmetry conserving configuration-mixing calculation of the generator coordinate method using a Skyrme [4] and RMF [5] density functionals and in terms of the collective Hamiltonian approach using a Gogny functional [6]. Same qualitative trend is observed in these kinds of studies, namely that the correlation energies are most enhanced for the transitional nuclei.

7.2 Empirical Proton-Neutron Correlation

We next consider the following quantity called δV_{pn}, which is essentially the double difference of the binding energy with respect to both proton and neutron number:

$$\delta V_{pn} = \frac{1}{4}[\{BE(Z, N) - BE(Z, N-2)\} - \{BE(Z-2, N) - BE(Z-2, N-2)\}], \tag{7.7}$$

where $BE(N, Z)$ stands for the binding energy, i.e., the minus of the total ground-state energy $-E$ for the nucleus with N and Z.

The δV_{pn} measures empirical interaction between last protons and neutrons, as investigated by Federmann and Pittel [7] and by Zhang et al. [8] in 1977. The empirical proton-neutron interaction gives insight into a microscopic origin of the onset of deformation [7], and hence the δV_{pn} is a stringent test for the shape transition. For a recent couple of years, Casten and his collaborators have extensively worked on the overall systematic trend of this quantity in connection with the quadrupole collectivity from an empirical point of view [9, 10]. The binding energy is supposed to be a sum of all interactions acting among nucleons. As the difference of the binding energy is nothing but the two-nucleon separation energies, the double difference can be used to estimate occurrence of collectivity, shell evolution and configuration mixing of different intrinsic shapes [7]. The general feature of the δV_{pn} is like this: When last protons and last neutrons occupy similar orbit, e.g., particle-particle (pp), where both protons and neutrons occupy the orbits from the beginning to the middle of the respective major shells, and hole-hole (hh) configurations, where both protons and neutrons surpass the middle of the major shells, then the overlap of the protons and the neutrons become large. However, in the case where the protons and neutrons occupy particle and hole orbits, or vise versa, the δV_{pn} then becomes relatively small. This situation is well illustrated in Fig. 7.4 for W, Os and Pt isotopes as representatives, and some general pattern is predicted in the right-lower quadrant of ^{208}Pb. The δV_{pn} is calculated in terms of the IBM Hamiltonian determined based on the constrained HF+BCS method with Skyrme SkM* functional as in the case of Fig. 7.3. For all the considered isotopes chains, the calculated δV_{pn} value in the upper panel of Fig. 7.4 looks relatively small being below 250 keV over the range from the beginning to the middle of the major shell $82 \leq N \leq 126$, where protons and neutrons are in hole and particle configurations, respectively. From the mid-shell $N = 106$ toward the end of the major shell $N = 126$, where both protons and neutrons are in the hole configurations, the δV_{pn} value becomes larger than 250 keV and keeps increasing. This calculated result is quite consistent with the experimental trend shown in the lower panel of Fig. 7.4. When going beyond the shell closure $N = 126$, the δV_{pn} value becomes much smaller as protons and neutrons occupy hole and particle orbits, respectively.

7.3 Brief Summary

Summarizing this chapter, some ground-state properties have been considered as an implication to the collective structural evolution of medium-heavy and heavy nuclei. When the IBM Hamiltonian is formulated by the microscopic energy density

7.3 Brief Summary

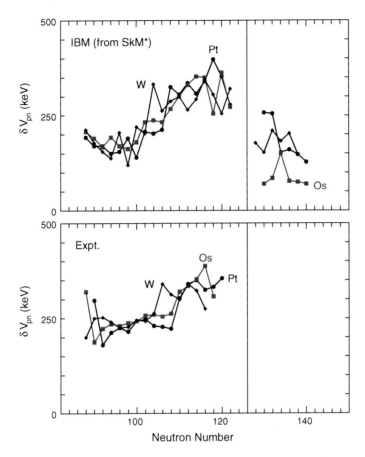

Fig. 7.4 Systematics of the δV_{pn} as function of the neutron number for W, Os and Pt nuclei as representatives. Skyrme SkM* functional is used

functional calculation and is diagonalized with good quantum numbers in the laboratory frame, then the binding energy obtained from the IBM Hamiltonian contains quantum-mechanical correlation effect which is otherwise not taken into account at the mean-field level. The effect of the correlation energy turned out be significant in reproducing the proper behaviour of the two-neutron separation energy of Sm isotopes reflecting the first-order shape-phase transition in Fig. 7.2. The robust feature that the correlation energy becomes most significant for the transitional regions has been examined in a systematic manner (cf. Fig. 7.3). In addition the quantity δV_{pn} has been introduced to provide a hint for the structural evolution concerning the relevant shell structure and the collectivity (cf. Fig. 7.4). It should be interesting in the future to extend the analysis shown in Fig. 7.4 over more wider range of isotopes involving exotic ones, and further to clarify which microscopic functional as well as which interaction of the IBM is most appropriate for extracting the knowledge about the nuclear structure relevant to the collectivity out of the δV_{pn} systematics.

References

1. Table of isotopes. Atomic mass data. http://ie.lbl.gov/
2. Scholten O, Iachello F, Arima A (1978) Interacting boson model of collective nuclear states III. The transition from SU(5) to SU(3). Ann Phys (NY) 115:325
3. Nomura K, Shimizu N, Otsuka T (2010) Formulating the interacting boson model by mean-field methods. Phys Rev C 81:044307
4. Bender M, Bertsch G, Heenen P-H (2006) Global study of quadrupole correlation effects. Phys Rev C 73:034322
5. Yao JM, Mei H, Chen H, Meng J, Ring P, Vretenar D (2011) Configuration mixing of angular-momentum-projected triaxial relativistic mean-field wave functions. II. Microscopic analysis of low-lying states in magnesium isotopes. Phys Rev C 83:014308
6. Delaroche J-P, Girod M, Libert L, Goutte H, Hilaire S, Peru S, Pillet N, Bertsch GF (2010) Structure of even-even nuclei using a mapped collective Hamiltonian and the D1S Gogny interaction. Phys Rev C 81:014303
7. Federmann P, Pittel S (1977) Towards a unified microscopic description of nuclear deformation. Phys Lett B 69:385
8. Zhang J-Y, Casten RF, Brenner DS (1989) Empirical proton-neutron interaction energies. Linearity and saturation phenomena. Phys Lett B 277:1
9. Cakirli RB, Casten RF (2006) Direct empirical correlation between proton-neutron interaction strengths and the growth of collectivity in nuclei. Phys Rev Lett 96:132501
10. Cakirli RB, Brenner DS, Casten RF, Millman EA (2005) Proton-neutron interactions and the new atomic masses. Phys Rev Lett 94:092501
11. Nomura K (2011) Microscopic derivation of IBM and structural evolution in nuclei, AIP Conference Proceedings 1355, "International Symposium; New Faces of Atomic Nuclei", 23–28

Chapter 8
Summary and Concluding Remarks

A novel and robust way of deriving the Hamiltonian of the interacting boson model (IBM) from the microscopic self-consistent mean-field theory has been introduced. The mean-field calculation employing a universal energy density functional (EDF) is a most reliable tool for describing the bulk properties of virtually all nuclei. The IBM on its own has made considerable success over the decades in reproducing low-lying collective structure of medium-heavy and heavy nuclei. In the present work the bridge was made over the gap between the two frameworks, which have been heretofore thought as being irrelevant to with each other. By mapping the mean-field to the appropriate IBM systems, the IBM Hamiltonian is derived, thereby allowing to calculate the energies and the wave functions of excited states keeping good quantum numbers in laboratory frame. As any state-of-the-art EDF is already well calibrated with the observed intrinsic properties of finite nuclei and is universal, one can derive the IBM Hamiltonian basically for all situations of the quadrupole collective dynamics including those far from β stability line, not yet studied experimentally. This is a great advantage in the era of third-generation rare isotope beams producing many new heavy exotic nuclei.

The strength parameters of the IBM Hamiltonian are determined by mapping each point on the nucleonic energy surface, obtained from the self-consistent mean-field calculation using a given microscopic EDF with the quadrupole constraint associated with the deformation parameters β and γ, onto an appropriate energy expectation value of the boson system in the coherent state. Through this process, all essential ingredients of the fermion many-body systems for the corresponding low-lying quadrupole collective states, namely the anti-symmetrization, nucleon-nucleon forces ..., can be simulated by the mathematically simpler boson model. In Chap. 2 we have described the way the mapping is performed by the application of the Wavelet analysis, as well as how well it works. By the Wavelet analysis can one determine the parameter values of the IBM in an unambiguous manner, which would be otherwise quite arbitrary when simply playing the parameters by hand. A problem arises when the model is applied to the strongly-deformed nuclei: the moment of inertia of the rotational band that is generated by the IBM Hamiltonian derived from the energy-surface mapping procedure turns out to be several tens per cent smaller than the observed moment of inertia.

On top of this fact, we have formulated a new fermion-to-boson mapping idea for the rotational motion of well-deformed axially symmetric nuclei in Chap. 3. It was found that, for strongly-deformed rotational nuclei, the nucleon system responds to the rotational cranking in a substantially different way from the boson system. This is a possible origin of the discrepancy of the IBM moment of inertia from the experimental data, and one should then go one step beyond to take the dynamical, non-zero frequency mode into account. In order that the bosonic response to the rotational cranking becomes identical to the nucleonic one, we proposed to introduce the rotational kinetic-like term known as the LL term in the phenomenology. Consequently, it turned out that the IBM is capable of describing the rotational moment of inertia microscopically and precisely at the quantitative level. The finding presented in Chap. 3 sheds light upon a criticism made in the past by Bohr and Mottelson that the IBM is far from sufficient to account for the intrinsic state of the deformed nuclei when formulated microscopically. The result provides one with a crucial piece of information as to the validity of the IBM on deformed nuclei, justifying the model for the first time in the case of the yrast states of axially-symmetric deformed nuclei.

In Chap. 4 we have studied spectroscopic properties of the medium-heavy nuclei basically with modest deformation, where the non-axial degrees of freedom plays a crucial role in determining the nuclear shape. The concepts of the quantum-phase transition and the critical point symmetries were reviewed, which help understanding the way the nuclear equilibrium shape changes as a function of the nucleon number. Several cases of the shape transitions were presented in comparison with the available experimental data: vibrational-to-γ-unstable transition in the mass $A \approx 100$–130 region, the prolate-to-oblate shape transition in $A \approx 180$–200 region. Heavy nuclei around the mass $A \approx 190$ exhibit competition between prolate and oblate intrinsic states, resulting in a spectacular shape coexistence as observed experimentally. We have studied the spectroscopic systematics of these nuclei based on the IBM Hamiltonian derived from finite-range Gogny EDF. Particularly we discussed the possibility of the coexistence of prolate and oblate shapes in Pt isotopes, and suggested that a single configuration, without introducing the intruder configuration of cross-shell excitation, is sufficiently well for Pt nuclei. We further predicted, prior to experimental studies, the transition from prolate to oblate shapes in heavy Os and W nuclei as a function of neutron number N and the transition points $N = 116$. The collective structural evolution was examined also for the neutron-rich exotic nuclei with which the experimental data have been so far quite sparse or completely missing. For the W and Os nuclei with $N > 126$, locating in the right-lower quadrant of the doubly-magic ^{208}Pb nucleus, the evidence for the E(5) critical-point symmetry was predicted in the wide rage of the nuclear chart. Along this line, we gave the first theoretical explanation on the γ-ray spectroscopy of neutron-rich Kr isotopes, carried out at the REX-ISOLDE facility at CERN. Contrary to some earlier measurements, we concluded from both theoretical and experimental viewpoints that the shape evolution in the considered Kr nuclei occurs quite slowly, and that no sudden onset of deformation is observed. This result certainly has a significant impact, because the neutron-rich nuclei with mass $A \approx 100$, including the Kr isotopes, represent an intersection of collective and single-particle degrees of freedom, and also should be of common interest for the studies of shell structure, QPT, mass measurement and astrophysical processes.

On the other hand, it should be quite curious to see whether the present model has an equal quality of predictive power to other spectra-generating EDF-based approach. The present model can be viewed as an alternative to the full-configuration mixing and/or the restoration of the broken symmetries, which are still much more involved and computationally demanding in the case where the triaxial degrees of freedom are included. In Chap. 5, we then compare the spectroscopic properties obtained from the IBM Hamiltonian derived by the relativistic DD-PC1 functional with those resulting from another approximation to the projected GCM configuration mixing calculation: the collective Hamiltonian approach based on the same DD-PC1 relativistic EDF, taking ^{192,194,196}Pt nuclei as examples. The ground-state band computed from both approaches were quite consistent with the available data. A marked difference can be seen in the ordering pattern of the quasi-γ band energies. It was shown that the quasi-γ band energies in IBM (composed of two-body boson terms, actually) form the couplets as $(3_\gamma^+, 4_\gamma^+)$, $(5_\gamma^+, 6_\gamma^+)$, ..., which is characteristic to the O(6) limit. The relevant excitation energies obtained from the collective Hamiltonian indicated, however, that the quasi-γ band energies are lying rather regularly, being much consistent with the experimental evidence.

This finding further sheds light upon another intriguing, but unsolved problem on the axially asymmetric nuclei. On top of the analysis mentioned above, a robust feature of the γ-soft nuclei and its microscopic description were presented in Chap. 6. It is well known empirically that many of triaxial nuclei fall in between the rigid triaxial-rotor model of Davydov-Filippov and the γ-unstable rotor model of Wilets-Jean. This fact was confirmed by introducing a specific form of the three-body boson Hamiltonian, which produces a shallow triaxial minimum which is seen also in the microscopic EDF calculation. The three-body Hamiltonian itself has been used in the phenomenological IBM-1 somewhat a priori, but here it was introduced in the microscopic IBM-2 framework. When describing the γ-soft nuclei, the earlier IBM descriptions have been connected too much with the O(6) nuclei, and the most axially asymmetric nuclei are thought as a perturbation to the O(6) dynamical symmetry. This work suggested, for the first time, that neither of the rigid-triaxial and the γ-unstable (or O(6) symmetry) rotor models is realized, from the perspective of the microscopic EDF approach. To measure the γ softness, not only the excitation energies, but the wave functions of the states in the quasi-γ band should be inspected. Such analysis is possible only by a microscopic analysis like the present work. What is more, the basic features present in Chap. 6 are sound and general as they do not depend on the choice of the microscopic energy density functionals.

As a supplementary study, nuclear quadrupole dynamics was investigated Chap. 7 from a slightly different perspective, in terms of the quantum-mechanical correlation effect on the ground state. In the transitional region between nearly spherical and deformed shapes, the quantum fluctuation becomes significant, and the correlation energy obtained from the diagonalization of the boson Hamiltonian was indeed shown to be maximal around the transitional nuclei in a given isotopic sequence. In the mean-field approximation, the discontinuity in the two-neutron separation energy of the first-order phase transition is rather unclear. In many cases of the mean-field calculations, an anomalous kink shows up at around the closed shell because shell effect is reflected too strongly. When the IBM is formulated microscopically, the resultant separation energy in Sm isotopes describes phase transition correctly. The present microscopic analysis can be applied even to exotic nuclei. Also the empirical proton-neutron interaction can be evaluated by the δV_{pn} plot, which reflects how the collectivity correlates with the underlying shell structure. These ground-state properties can be used for quantifying another novel shape dynamics from a microscopic picture.

Major outcomes resulting from the present work are the following:

- Bridge has been made over the gap between nuclear EDF and IBM. Their complementarity can be utilized, thereby allowing to derive the spectroscopic properties with good quantum numbers and to take the correlation effect into account. Main part of the two-body IBM Hamiltonian is determined by the mapping from the microscopic EDF energy surface. This works well basically for vibrational and γ-soft systems with relatively weak deformation.
- In the case of strongly deformed rotational nuclei, one has to go one step forward. To describe the moment of inertia of rotational band correctly, a kinetic-like LL term becomes necessary. As the LL term does not change the wave function, the main part of the IBM Hamiltonian can be kept.
- If the notable triaxiality enters, Hamiltonian with up to two-body boson terms does not suffice and three-body boson term should be introduced. With a suitably chosen three-body Hamiltonian, empirical regularity of non-axial nuclei can be explained naturally.

Possible future directions are as follows:

- First, the finding of this thesis naturally points to the interest to the microscopic EDF itself and indicates that one should further investigate its validity. Since most of the highly-reputed EDFs are constructed from the bulk properties of stable nuclei, some spectroscopic properties for well-deformed nuclei cannot be always described well by the present methodology. It is not completely clear whether a conventional EDF still has an enough predictive power for exotic nuclei near the drip line where much complex correlation should enter. In contrast to the original density-functional approach in quantum chemistry, the EDF for nuclear many-body system is not unique, and the modeling of universal EDF should be an open question. Moreover, some important piece of nuclear effective interaction like the tensor force still remains to be included in the EDF-based mean-field models. Also it should be quite interesting to address methodological problems encountered in the restoration of symmetries broken in the mean-field approximation.
- The second possibility which would certainly attract considerable broad interest is to apply the methodology presented in this thesis to other fields of physics such as molecular and atomic systems, and other mesoscopic quantum many-body

systems. In these systems, spectacular rotational and vibrational spectra, as well as the phase transitions, similar to those in nuclear physics are observed. Also the similar kind of the algebraic model to the IBM for atomic nucleus exists in the mesoscopic system (for instance, Ref. [1] and for reviews, see Refs. [2, 3] and references therein), while the reliable *ab initio* description of it is provided by density functional theory, Hartree-Fock method, etc. Both of these macroscopic (algebraic) and microscopic (density-functional or HF) theories could be merged as was done here, since in chemistry an energy landscape, similar to the one for the nuclear intrinsic deformation, can be defined in terms of the geometrical configuration of constituents. More specifically, the isovector collective motion of two-fluid quantum system, e.g., the scissors mode in axially deformed nuclei (for thorough review, see Ref. [4] and references are therein), is a general phenomenon, which is not limited to the nuclear physics but can be seen in other physical systems including the trapped Bose-Einstein condensate [5]. Then the concept of the proton-neutron mixed-symmetry state in the IBM-2 (see, e.g., Ref. [6] and references are therein) would be applied to these intriguing subjects. These ideas are brand new, and will bring about an important theoretical advancement in the understanding of general quantum many-body systems.

These issues will be under thorough investigation in the future.

References

1. Iachello F (1981) Algebraic methods for molecular rotation-vibration spectra. Chem Phys Lett 78:581
2. Frank A, Van Isacker P (1994) Algebraic methods in molecular and nuclear structure physics. Willey, New York
3. Iachello F, Levine RD (1995) Algebraic theory of molecules. Oxford University Press, Oxford
4. Heyde K, von Brentano P, Richter A (2010) Magnetic dipole excitations in nuclei: elementary modes of nucleonic motion. Rev Mod Phys 82:2365
5. Maragò OM, Hopkins SA, Arlt J, Hodby E, Hechenblaikner G, and Foot CJ (2000) Observation of the scissors mode and evidence for superfluidity of a trapped Bose-Einstein condensed gas. Phys Rev Lett 84:2056
6. Pietralla N, von Brentano P, Lisetskiy AF (2008) Experiments on multiphonon states with proton-neutron mixed symmetry in vibrational nuclei. Prog Part Nucl Phys 60:225

Appendix A
Details of Mean-Field Calculations

In this chapter, we shall give a brief explanation of the self-consistent mean-field model. The following description is based on Refs. [1–3].

A.1 Hartree-Fock Method

In the self-consistent mean-field model, a single particle is assumed to be in the average field created by all other surrounding particles, and one should start with a trial wave function comprised of all these particles. The trial wave function normally takes the form of single Slater determinant, which is called Hartree-Fock basis and is composed of A single-particle wave function under the averaged field. The Slater determinant is thus given as

$$|\Phi\rangle = a_1^\dagger a_2^\dagger \cdots a_A^\dagger |-\rangle. \qquad (A.1)$$

where $|-\rangle$ and $a_k^\dagger$ being the (bare) vacuum and a creation operator for a single-particle wave function ϕ_k, respectively. A Schrödinger equation for a single-particle wave function is written as $h_k \phi_k = \varepsilon_k \phi_k$, where h_k and ε_k stand for the Hamiltonian for k-th particle and its eigenvalue, respectively. In general, ϕ_k can be expanded by using the complete and orthonormal basis set, $\{\chi_l\}$ as

$$\phi_k = \sum_l D_{lk} \chi_l \qquad (A.2)$$

where D_{lk} is a unitary operator and χ's should correspond to some creation (annihilation) operator $c_k^{(\dagger)}$. $\{\chi\}$ can be any basis set that is solved already, e.g., the harmonic oscillator basis, etc. Similarly to the single-particle wave functions,

$$a_k^\dagger = \sum_l D_{lk} c_l^\dagger \quad (A.3)$$

is fulfilled. The expectation value of the general many-body Hamiltonian $\hat{H}$ with up to two-body interactions, $\langle \Phi | \hat{H} | \Phi \rangle$, is minimized in terms of the Slater determinant $|\Phi\rangle$. The optimized single-particle properties and total energies of given nuclei are obtained iteratively until a good convergence is achieved. It is the outline of the Hartree-Fock method.

Instead of D's, it is more convenient to introduce the density matrix as

$$\rho_{\alpha\beta} = \langle \Phi | c_\beta^\dagger c_\alpha | \Phi \rangle \quad (A.4)$$

which can be written in terms of $a^\dagger$ operator as

$$\rho_{\alpha\beta} = \sum_{\gamma\delta} D_{\alpha\gamma} D_{\delta\beta}^* \langle \Phi | a_\beta^\dagger a_\alpha | \Phi \rangle = \sum_{i=1}^{A} D_{\alpha i} D_{i\beta}^* \quad (A.5)$$

A single-particle density matrix is directly related to the Slater determinant. In fact, it only has eigenvalues 0 and 1,[1] corresponding to particle and hole states, respectively. Since a small variation $\rho + \delta\rho$ is also a projector,

$$(\rho + \delta\rho)^2 = \rho + \delta\rho \quad (A.6)$$

which, up to the terms linear in $\delta\rho$, leads us to

$$\delta\rho = \rho\delta\rho + \delta\rho\rho. \quad (A.7)$$

As ρ is diagonal in the Hartree-Fock basis, the particle-particle and hole-hole matrix elements of $\delta\rho$ have to vanish

$$\rho\delta\rho\rho = \sigma\delta\rho\sigma = 0, \quad (A.8)$$

where $\sigma (=1-\rho)$ is also a projection operator[2] onto the subspace spanned by particle states.

Let us consider the many-body Hamiltonian with up to two-body interactions

$$\hat{H} = \sum_{\alpha\beta} t_{\alpha\beta} c_\alpha^\dagger c_\beta + \frac{1}{4} \sum_{\alpha\beta\gamma\delta} \bar{v}_{\alpha\beta\gamma\delta} c_\alpha^\dagger c_\beta^\dagger c_\delta c_\gamma, \quad (A.9)$$

---

[1] Since ρ is a projector, $\rho^2 = \rho$ is satisfied.
[2] In fact, $\sigma^2 = \sigma$ is satisfied.

Appendix A: Details of Mean-Field Calculations

with $t_{\alpha\beta}$ and $\bar{v}_{\alpha\beta\gamma\delta}$ being the matrix elements for one- and two-body interactions, respectively, where the latter is anti-symmetrized: in terms of γ and δ as

$$\bar{v}_{\alpha\beta\gamma\delta} = v_{\alpha\beta\gamma\delta} - v_{\alpha\beta\delta\gamma}. \tag{A.10}$$

The energy density functional for many-body Hamiltonian is expressed as

$$E[\rho] = \sum_{\alpha\beta} t_{\alpha\beta} \langle \Phi | c_\beta^\dagger c_\alpha | \Phi \rangle + \frac{1}{4} \sum_{\alpha\beta\gamma\delta} \bar{v}_{\alpha\beta\gamma\delta} \langle \Phi | c_\alpha^\dagger c_\beta^\dagger c_\delta c_\gamma | \Phi \rangle \tag{A.11}$$

$$= \sum_{\alpha\beta} t_{\alpha\beta} \rho_{\alpha\beta} + \frac{1}{2} \sum_{\alpha\beta\gamma\delta} \rho_{\gamma\alpha} \bar{v}_{\alpha\beta\gamma\delta} \rho_{\delta\beta}.$$

where the Wick's theorem is used for $c^{(\dagger)}$'s.

The variation with respective to the density matrix leads one to

$$\delta E = E[\rho + \delta\rho] - E[\rho] = \sum_{\alpha\beta} h_{\alpha\beta} \delta\rho_{\alpha\beta} = \sum_{mi} (h_{mi}\delta\rho_{im} + h_{im}\delta\rho_{mi}), \tag{A.12}$$

where m and i represent particle and hole occupying the orbits above and below the Fermi surface, respectively. Hamiltonian density matrix $h_{\alpha\beta}$ can be defined as

$$h_{\alpha\beta} = \frac{\partial E[\rho]}{\partial \rho_{\beta\alpha}} = t_{\alpha\beta} + \Gamma_{\alpha\beta}, \tag{A.13}$$

where $\Gamma_{\alpha\beta} = \sum_{\gamma\delta} \bar{v}_{\alpha\delta\beta\gamma} \rho_{\alpha\beta}$ is the self-consistent field. The particle-hole elements of the Hamiltonian density matrix have to vanish. h and ρ are diagonalized simultaneously,

$$[h, \rho] = 0 \tag{A.14}$$

which is nothing but the Hartree-Fock equation in terms of the density matrix. We get the Hartree-Fock equation

$$h_{\alpha\beta} = t_{\alpha\beta} + \sum_{i=1}^{A} \bar{v}_{\alpha i \beta i} = \varepsilon_\alpha \delta_{\alpha\beta} \tag{A.15}$$

In principle, the Hartree-Fock equation is nonlinear with respect to $\rho_{\rho\beta}$, so that it is solved iteratively until its solution is converged to certain value. The single-particle energy can be obtained from the eigenvalue problem. The total energy in this case is a sum of the single-particle energy. On the other hand, it is sometimes useful to solve the HF equation by employing the 3D Cartesian grid like in the present study.

A.2 Hartree-Fock Plus BCS Calculation

The effect of the pairing correlation should be often considered for nuclear system. Among many types of pairing models, we here describe the BCS approximation and how it is applied to the Hartree-Fock theory. A particle and the time-reversal conjugate of the particle are coupled with angular momentum zero and stay in the same orbit k. The following BCS state[3] is introduced independently for proton and neutron in the same way as the general theory of superconductivity [4, 5],

$$|\text{BCS}\rangle = \prod_{k>0}(u_k + v_k a_k^\dagger a_{\bar{k}}^\dagger)|-\rangle, \tag{A.16}$$

where the single-particle state k has been determined by the preceding Hartree-Fock calculations and the state $\bar{k}$ represents the time reversal conjugate of the state k. The real parameters u_k and v_k are variational parameters, related to the occupation probabilities of particles satisfying

$$u_k^2 + v_k^2 = 1. \tag{A.17}$$

The meanings of the parameters v_k^2 and u_k^2 are the probabilities the paired particles occupy and do not occupy the orbit k, respectively. In the framework of Hartree-Fock plus BCS calculation, variations are carried out over the single-particle wave function and the factors (u_k, v_k) separately.[4] Since the trial wave function, i.e., BCS ground state Eq. (A.16), is not the eigenvalue of the particle number, a subsidiary condition as to the particle number with a Lagrange multiplier should be imposed as

$$\hat{H}' = \hat{H} - \lambda \hat{N}, \tag{A.18}$$

where $\hat{H}$ and $\hat{N}$ are many-body Hamiltonian and the particle-number operator, respectively. We assume that the former is represented as

$$\hat{H} = \sum_k t_k a_k^\dagger a_k - G \sum_{k,k'>0} . \tag{A.19}$$

On the other hand, the number operator $\hat{N}$ is defined as

$$\hat{N} = \sum_{k>0}(a_k^\dagger a_k + a_{\bar{k}}^\dagger a_{\bar{k}}), \tag{A.20}$$

which in fact is a desired number of proton or neutron

---

[3] The normalization of the BCS state $\langle \text{BCS}|\text{BCS}\rangle = 1$ is guaranteed.
[4] In the Hartree-Fock-Bogoliubov (HFB) method, a generalized theory of the HF+BCS, quasi particles are introduced and the variations over the wave functions and the occupation factors are carried out simultaneously.

Appendix A: Details of Mean-Field Calculations

$$\langle \text{BCS}|\hat{N}|\text{BCS}\rangle = 2\sum_{k>0} v_k^2 = N. \tag{A.21}$$

Although, in the HF+BCS calculations presented in this thesis, we employ the density-dependent δ-function type of the pairing force, we here discuss the case of a pure pairing force with a constant matrix element $-G$ to make the discussion as simple as possible. In such case, the expectation value of the Hamiltonian $\hat{H}'$ in Eq. (A.18) can be evaluated by the variational condition with respect to u_k and v_k

$$0 = \delta\langle\text{BCS}|\left(\hat{H} - \lambda\hat{N}\right)|\text{BCS}\rangle \tag{A.22}$$
$$= \frac{\partial}{\partial v_k}\langle\text{BCS}|\left\{\sum_k (t_k - \lambda)a_k^\dagger a_k - G\sum_{kk'>0} a_k^\dagger a_{\bar{k}}^\dagger a_{\bar{k}'}a_{k'}\right\}|\text{BCS}\rangle,$$

where the derivative in terms of v_k

$$\frac{\partial}{\partial v_k} = \frac{\partial}{\partial v_k}\bigg|_{u_k} - \frac{v_k}{u_k}\frac{\partial}{\partial u_k}\bigg|_{v_k} \tag{A.23}$$

is satisfied through $u_k^2 + v_k^2 = 1$.

By using the following matrix elements of $a_k^\dagger a_k$ and $a_k^\dagger a_{\bar{k}}^\dagger a_{\bar{k}'} a_{k'}$,

$$\langle\text{BCS}|a_k^\dagger a_k|\text{BCS}\rangle = v_k^2 \quad \text{and} \tag{A.24}$$
$$\langle\text{BCS}|a_k^\dagger a_{\bar{k}}^\dagger a_{\bar{k}'} a_{k'}|\text{BCS}\rangle = u_k v_k u_{k'} v_{k'} \quad (\text{for } k \neq k')$$
$$= v_k^2 \quad (\text{for } k = k'),$$

Equation (A.22) can be expressed as

$$2\varepsilon_k v_k u_k + \Delta(v_k^2 - u_k^2) = 0, \tag{A.25}$$

where notations of $\Delta \equiv G\sum_{k>0} u_k v_k$ and $\varepsilon_k \equiv t_k - \lambda - Gv_k^2$ are introduced. From Eqs. (A.25) and (A.17), v_k^2 and u_k^2 can be expressed by ε_k and Δ as

$$v_k^2 = \frac{1}{2}\left(1 - \frac{\varepsilon_k}{\sqrt{\varepsilon_k^2 + \Delta^2}}\right), \quad u_k^2 = \frac{1}{2}\left(1 + \frac{\varepsilon_k}{\sqrt{\varepsilon_k^2 + \Delta^2}}\right) \tag{A.26}$$

Substituting the Eq. (A.26) into the definition of Δ, one obtains so-called gap equation

$$\Delta = \frac{G}{2} \sum_{k>0} \frac{\Delta}{\sqrt{\varepsilon_k^2 + \Delta^2}}, \qquad (A.27)$$

which can be solved iteratively using the known value of G.

A.3 Constrained Hartree-Fock Calculation

In the Hartree-Fock (with pairing correlation being included) theory, the ground-state energy of a given system is obtained by minimizing the energy functional with respect to the density. Nevertheless, the HF calculation gives rise to only local minimum, while whole energy landscape in terms of some coordinates becomes quite important particularly for the discussion on the surface deformation, fission dynamics, etc.

One needs to know the wave function $\Phi(q)$ which optimizes the total energy under the constraint that a certain operator on a single particle state has a fixed expectation value $q = \langle \Phi | \hat{Q} | \Phi \rangle$, where $\langle \hat{Q} \rangle$ is multipole external field which takes places the intrinsic deformation. A simple way is to impose the linear constraint with a Lagrange multiplier λ to the Hamiltonian and minimizing the $\langle \hat{H}' \rangle = \langle \hat{H} \rangle - \lambda \langle \hat{Q} \rangle$, which is similar to the treatment of the particle number in the BCS approximation. Linear constraint is only applicable when the potential energy curve has a positive second derivative. In this case, the variation with respect to q may no longer yield a stable solution.

Hartree-Fock calculations with quadratic constraint are carried out in such a way that the total Hamiltonian, consisting of unconstrained Hamiltonian plus the parabola $\frac{1}{2}C(\langle Q \rangle - \mu)^2$ at $\mu = q$, is evaluated by the variational principle. This yields the equation $\delta \langle \hat{H} \rangle - C(\mu - \langle \hat{Q} \rangle) \delta \langle \hat{Q} \rangle = 0$, which is equivalent to a linear constraint with $\lambda = C(\mu - \langle \hat{Q} \rangle)$. The value of λ is revised at each step of the iteration. The energy landscape obtained by the quadratic constraint is the same as the one by linear constraint.

Appendix B
Formulas in the IBM-2 Framework

This chapter presents some formulae in the IBM calculation. We first describe the coherent-state framework in Sect. B.1 and explain the M-scheme diagonalization in Sect. B.2. For further details of the coherent-state framework, the reader is referred to Refs. [7–9].

B.1 Coherent-State Formalism

The IBM-2 Hamiltonian With up to Two-Body Terms

The expectation value of the boson Hamiltonian $\hat{H}_{\text{IBM}}$ of Eq. (2.41), $\langle \hat{H}_{\text{IBM}} \rangle$ in the coherent state $|\Phi(N_\pi, N_\nu, \beta_B, \gamma_B)\rangle$ in Eq. (2.36) is calculated.

To this end the following general formula for any operators $\hat{f}$ and $\hat{g}$ can be used:

$$[\hat{f}, \hat{g}^N] = N\hat{g}^{N-1}[\hat{f}, \hat{g}] + \frac{1}{2}N(N-1)\hat{g}^{N-2}[[\hat{f}, \hat{g}], \hat{g}] \quad (\text{B.1})$$
$$+ \frac{1}{6}N(N-1)(N-2)\hat{g}^{N-3}[[[\hat{f}, \hat{g}], \hat{g}], \hat{g}] + \cdots,$$

where [,] indicates as usual the commutation relation.

We recall the Hamiltonian containing all the interaction terms considered in this thesis:

$$\hat{H}_{\text{IBM}} = \varepsilon_\pi \hat{n}_{d\pi} + \varepsilon_\nu \hat{n}_{d\nu} + \kappa \hat{Q}_\pi \cdot \hat{Q}_\nu + \alpha \hat{L} \cdot \hat{L}. \quad (\text{B.2})$$

The expectation value of the d-boson number operator on the RHS of Eq. (B.2):

$$\varepsilon_\rho \langle \hat{n}_\rho \rangle = \frac{N_\rho}{1+\beta_\rho} \varepsilon_\rho \left(\frac{1}{2}\beta_\rho^2 \sin^2 \gamma_\rho + \beta_\rho^2 \cos^2 \gamma_\rho + \frac{1}{2}\beta_\rho^2 \sin^2 \gamma_\rho \right) \quad (\text{B.3})$$

$$= N_\rho \varepsilon_\rho \frac{\beta_\rho^2}{1+\beta_\rho^2}.$$

For the quadrupole-quadrupole interaction between proton and neutron systems,

$$\kappa \langle \hat{Q}_\pi \cdot \hat{Q}_\nu \rangle = \frac{\kappa N_\pi N_\nu}{(1+\beta_\pi^2)(1+\beta_\nu^2)} \Big[4\beta_\pi \beta_\nu \cos(\gamma_\pi - \gamma_\nu) \qquad (B.4)$$

$$- 2\sqrt{\frac{2}{7}} \chi_\pi \beta_\nu \beta_\pi^2 \cos(\gamma_\nu + 2\gamma_\pi) - 2\sqrt{\frac{2}{7}} \chi_\nu \beta_\pi \beta_\nu^2 \cos(\gamma_\pi + 2\gamma_\nu)$$

$$+ \frac{2}{7} \chi_\pi \chi_\nu \beta_\pi^2 \beta_\nu^2 \cos 2(\gamma_\pi - \gamma_\nu) \Big].$$

The expectation value of the LL term reads

$$\alpha \langle \hat{L} \cdot \hat{L} \rangle = \alpha \langle \hat{L}_\pi \cdot \hat{L}_\pi + \hat{L}_\nu \cdot \hat{L}_\nu \rangle \qquad (B.5)$$

$$= \alpha \sum_{\rho=\pi,\nu} \frac{6 N_\rho \beta_\rho^2}{1+\beta_\rho^2}.$$

Note that $\langle \hat{L} \cdot \hat{L} \rangle$ differs from $\langle \hat{n}_d \rangle$ by the factor 6α. Then the factor 6α is renormalized into the d-boson energy ε.

From Eqs. (B.3)–(B.5), assuming that $\beta_\pi = \beta_\nu = \beta_B (= C_\beta \beta_F)$ and $\gamma_\pi = \gamma_\nu = \gamma_B (= \gamma_F)$ in Eq. (2.35), the most basic form of the IBM energy surface is calculated as

$$E(\beta_B, \gamma_B) = \frac{\varepsilon (N_\pi + N_\nu) \beta_B^2}{1+\beta_B^2} + \kappa N_\pi N_\nu \qquad (B.6)$$

$$\times \frac{\beta_B^2}{(1+\beta_B^2)^2} \Big[4 - 2\sqrt{\frac{2}{7}} (\chi_\pi + \chi_\nu) \beta_B \cos 3\gamma_B + \frac{2}{7} \chi_\pi \chi_\nu \beta_B^2 \Big],$$

which is identical to the expression in Eq. (2.46). With the above assumption of the deformation parameters, all Majorana terms vanish.

When $\hat{Q}_\rho \cdot \hat{Q}_\rho$ ($\rho = \pi$ or ν) terms are included, the expectation value of them can be calculated:

$$\kappa_\rho \langle Q_\rho \cdot Q_\rho \rangle = \kappa_\rho \Bigg\{ \frac{n_\rho}{1+\beta_\rho^2} [5 + (1+\chi_\rho^2)\beta_\rho^2] \qquad (B.7)$$

$$+ \frac{N_\rho (N_\rho - 1)\beta_\rho^2}{(1+\beta_\rho^2)^2} \Big[4 - 4\sqrt{\frac{2}{7}} \chi_\rho \beta_\rho \cos 3\gamma_\rho + \frac{2}{7} \chi_\rho^2 \beta_\rho^2 \Big] \Bigg\},$$

where κ_ρ is the intensity of the interaction. This term could be in general included, but is not used in practice because otherwise the number of parameters would increase.

Three-Body (Cubic) Boson Terms

When the triaxiality should be taken into account, the three-body (so-called cubic) boson term becomes important as it creates a stable non-axial minimum in the corresponding energy surface. In the case of IBM-1, such three-body term has been considered in the context of the phenomenological study. In this thesis, we introduced the three-body term in the microscopic IBM-2 framework. As this is entirely new, we need to give a detailed explanation here.

Given the proton-neutron degrees of freedom, one has to consider the three possibilities concerning the combination of proton-neutron bosons: two of them are obviously the term comprised of three like bosons, while the other is mixing of proton and neutron bosons. For medium-heavy and heavy nuclei, proton-neutron interaction dominates over the proton-proton and neutron-neutron ones. Thus, the former type, that is, the terms consisting of either proton or neutron bosons only, is less relevant than the latter type containing both proton and neutron bosons.

Also the cubic term can be composed of only d bosons in order to create the triaxial minimum in the energy surface. The most general form of the three-body term is given (cf. Eq. (6.2) in Chap. 6) as

$$\hat{H}_{3B} = \sum_{\rho' \neq \rho} \sum_L \theta_L^\rho [d_\rho^\dagger d_\rho^\dagger d_{\rho'}^\dagger]^{(L)} \cdot [\tilde{d}_{\rho'} \tilde{d}_\rho \tilde{d}_\rho]^{(L)}. \tag{B.8}$$

Among the possibilities of $L = 0, 2, 3, 4, 6$, only the $L = 3$ component gives rise to the stable triaxial minimum at $\gamma = 30°$. Then, other L components could be neglected to a first good approximation. Note that the ordering of the proton and neutron d bosons are of less importance, as indeed they commute. Of the $L = 3$ terms, there are two independent combinations for the intermediate coupling of the three bosons: Two bosons are coupled first with intermediate angular momentum $L' = 2$ and 4. In this thesis we only consider the intermediate angular momentum of $L' = 2$ for the sake of simplicity. Therefore, the actual three-body term used in this thesis is written more explicitly as

$$\hat{H}_{3B} = \theta_3 \sum_{\rho' \neq \rho} [[d_\rho^\dagger d_\rho^\dagger]^{(2)} d_{\rho'}^\dagger]^{(3)} \cdot [\tilde{d}_{\rho'} [\tilde{d}_\rho \tilde{d}_\rho]^{(2)}]^{(3)}, \tag{B.9}$$

where the coefficient θ_3^ρ is assumed to be identical between proton and neutron bosons, again for simplicity. The proton-proton-neutron part of Eq. (B.9) is given as

$$\hat{H}_{3B} = [[d_\pi^\dagger \times d_\pi^\dagger]^{(2)} \times d_\nu^\dagger]^{(3)} \cdot [[\tilde{d}_\pi \times \tilde{d}_\pi]^{(2)} \times \tilde{d}_\nu]^{(3)} \tag{B.10}$$
$$= \sum_\mu [[d_\pi^\dagger \times d_\pi^\dagger]^{(2)} \times d_\nu^\dagger]_\mu^{(3)} [[\tilde{d}_\pi \times \tilde{d}_\pi]^{(2)} \times \tilde{d}_\nu]_\mu^{(3)}$$

$$= \sum_{\mu,\mu_1,\mu_2,\mu_3,\mu_4} (22\mu_1\mu - \mu_1|3\mu)(22\mu_2\mu_1 - \mu_2|2\mu_1)(22\mu_3 - \mu - \mu_3|3 - \mu)$$

$$\times (22\mu_4\mu_3 - \mu_4|2\mu_3) d_{\pi,\mu_2}^\dagger d_{\pi,\mu_1-\mu_2}^\dagger d_{\pi,-\mu_4} d_{\pi,\mu_4-\mu_3} d_{\nu,\mu-\mu_1}^\dagger d_{\nu,\mu+\mu_3}$$

The two-body neutron boson part of the coherent-state expectation value of $\langle \hat{H}_{3B} \rangle$ is then given as

$$\frac{1}{1+\beta_\nu^2} \langle 0|(\lambda_\nu)^{N_\nu} d_{\nu,\mu-\mu_1}^\dagger d_{\nu,\mu+\mu_3} (\lambda_\nu^\dagger)^{N_\nu} |0\rangle. \tag{B.11}$$

Utilizing the general formula of Eq. (B.1), one obtains

$$[d_{\nu,\mu+\mu_3}, \lambda_\nu^\dagger] = a_{\nu,\mu+\mu_3} \quad \text{and} \quad [d_{\nu,\mu-\mu_1}^\dagger, \lambda_\nu] = -a_{\nu,\mu-\mu_1}, \tag{B.12}$$

which lead to

$$[d_{\nu,\mu+\mu_3}, (\lambda_\nu^\dagger)^{N_\nu}] = N_\nu a_{\nu,\mu+\mu_3} (\lambda_\nu^\dagger)^{N_\nu-1} \tag{B.13}$$

and

$$[d_{\nu,\mu-\mu_1}^\dagger, (\lambda_\nu)^{N_\nu}] = -N_\nu a_{\nu,\mu-\mu_1} (\lambda_\nu)^{N_\nu-1}, \tag{B.14}$$

respectively. Then, the expectation value for the neutron part is calculated:

$$\frac{N_\nu a_{\nu,\mu+\mu_3} a_{\nu,\mu-\mu_1}}{1+\beta_\nu^2} (22\mu_1\mu - \mu_1|3\mu)(22\mu_3 - \mu - \mu_3|3 - \mu). \tag{B.15}$$

Likewise, the proton part is written in the following form:

$$\frac{N_\pi(N_\pi - 1) a_{\pi,\mu_2} a_{\pi,\mu_4-\mu_3} a_{\pi,\mu_1-\mu_2} a_{\pi,-\mu_4}}{(1+\beta_\pi^2)^2} (22\mu_2\mu_1 - \mu_2|2\mu_1)(22\mu_4\mu_3 - \mu_4|2\mu_3). \tag{B.16}$$

The expectation value of the cubic term is obtained:

$$\langle \hat{H}_{3B} \rangle = \frac{N_\pi(N_\pi - 1) N_\nu}{(1+\beta_\nu^2)(1+\beta_\pi^2)^2} \tag{B.17}$$

$$\sum_{\mu,\mu_1,\mu_2,\mu_3,\mu_4} a_{\pi,\mu_2} a_{\pi,\mu_4-\mu_3} a_{\pi,\mu_1-\mu_2} a_{\pi,-\mu_4} a_{\nu,\mu+\mu_3} a_{\nu,\mu-\mu_1} (22\mu_1\mu - \mu_1|3\mu)$$

$$\times (22\mu_3 - \mu - \mu_3|3 - \mu)(22\mu_2\mu_1 - \mu_2|2\mu_1)(22\mu_4\mu_3 - \mu_4|2\mu_3).$$

If the equalities $\beta_\pi = \beta_\nu$ and $\gamma_\pi = \gamma_\nu$ are assumed and if one takes into account the $\hat{H}_{\pi\nu\nu}^{(3)}$ term, the cubic terms that contribute to the potential energy surface are written as

$$\langle \hat{H}_{\pi\pi\nu}^{(3)} + \hat{H}_{\pi\nu\nu}^{(3)} \rangle = -\frac{1}{7} N_\pi N_\nu (N_\pi + N_\nu - 2) \frac{\beta^6}{(1+\beta^2)^3} \sin^2 3\gamma. \tag{B.18}$$

Appendix B: Formulas in the IBM-2 Framework

Rotation of Intrinsic State

The overlap of the intrinsic wave functions of rotated and non-rotated states can be written as

$$\langle \Phi | e^{-i\theta \hat{L}_x} | \Phi \rangle = \langle 0 | \prod_{\rho=\pi,\nu} b_\rho^{N_\rho} e^{-i\theta \hat{L}_x} (b_\rho^\dagger)^{N_\rho} | 0 \rangle \tag{B.19}$$

$$= \langle 0 | b_\pi^{N_\pi} e^{-i\theta \hat{L}_{\pi x}} (b_\pi^\dagger)^{N_\pi} | 0 \rangle \langle 0 | b_\nu^{N_\nu} e^{-i\theta \hat{L}_{\nu x}} (b_\nu^\dagger)^{N_\nu} | 0 \rangle$$

$$= \langle 0 | b_\pi^{N_\pi} (e^{-i\theta \hat{L}_{\pi x}} b_\pi^\dagger e^{i\theta \hat{L}_{\pi x}})^{N_\pi} | 0 \rangle \langle 0 | b_\nu^{N_\nu} (e^{-i\theta \hat{L}_{\nu x}} b_\nu^\dagger e^{-i\theta \hat{L}_{\nu x}})^{N_\nu} | 0 \rangle$$

$$= \langle 0 | b_\pi^{N_\pi} (\lambda_\pi^\dagger)^{N_\pi} | 0 \rangle \langle 0 | b_\nu^{N_\nu} (\lambda_\nu^\dagger)^{N_\nu} | 0 \rangle$$

with $\hat{L}_{\rho x}$ being $\hat{L}_{\rho x} = (\hat{L}_{\rho -1} - \hat{L}_{\rho +1})/\sqrt{2}$. Since the present discussion is focused on axial symmetry, the boson coherent state is reduced to $|\Phi\rangle = \prod_{\rho=\pi,\nu} (s_\rho^\dagger + \beta_\rho d_{\rho 0}^\dagger)^{N_\rho} |0\rangle$. Here $\lambda_\rho^\dagger$ is defined as

$$\lambda_\rho^\dagger = e^{-i\theta \hat{L}_{\rho x}} b_\rho^\dagger e^{+i\theta \hat{L}_{\rho x}}. \tag{B.20}$$

Using the Baker-Hausdorff lemma

$$e^{x\hat{A}} \hat{B} e^{-x\hat{A}} = \hat{B} + x[\hat{A}, \hat{B}] + \frac{1}{2!} x^2 [\hat{A}, [\hat{A}, \hat{B}]] + \frac{1}{3!} x^3 [\hat{A}, [\hat{A}, [\hat{A}, \hat{B}]]] + \cdots \tag{B.21}$$

with $\hat{A}$ and $\hat{B}$ being any operators. Because $\lambda_\rho^\dagger$ for proton and neutron bosons can be calculated separately, the index ρ is omitted hereafter.

$$\langle \Phi | e^{-i\theta \hat{L}_x} | \Phi \rangle = N! (1 + \beta^2 d_{00}^2(\theta))^N, \tag{B.22}$$

where d_{00}^2 is the Wigner's d-function of the rotation. Taking into account the proton and neutron degrees of freedom, we have

$$\langle \Phi | e^{-i\theta \hat{L}_x} | \Phi \rangle = N_\pi ! N_\nu ! \left\{ 1 + \frac{\beta^2}{2} (3\cos^2 \theta - 1) \right\}^{N_\pi + N_\nu}. \tag{B.23}$$

The overlap of Eq. (B.24) should be normalized so as to be unit at $\theta = 0$.

$$\langle \Phi | e^{-i\theta \hat{L}_x} | \Phi \rangle = \left\{ \frac{1 + \beta^2 (3\cos^2 \theta - 1)/2}{1 + \beta^2} \right\}^{N_\pi + N_\nu} \tag{B.24}$$

Cranking Formula

The moment of inertia of the IBM system is obtained from the cranking formula of Ref. [6]. Since the time-reversal invariance is broken in the cranking formalism, one has to include the $d_{\pm 1}$ bosons in the coherent state of Eq. (2.36). Thus the energy expectation value $E = \langle \hat{H}_{\text{IBM}} \rangle$ is specified not only by $a_{\rho \pm 2}$ and $a_{\rho 0}$ but also by $a_{\rho \pm 1}$ components. Around the stationary point, the amplitudes $a_{\rho \mu}$ and the expectation values of the IBM Hamiltonian $\langle \hat{H}_{\text{IBM}} \rangle$ and the x-component of the angular momentum operator $L_x = \langle \hat{L}_x \rangle = \langle \hat{L}_{\pi x} + \hat{L}_{\nu x} \rangle$ are expanded as

$$a_{\rho\mu} = a_{\rho\mu}^{(0)} + \Delta_{\rho\mu}, \tag{B.25}$$

$$E = E^{(0)} + \sum_{\rho,\rho'} \sum_{\mu,\lambda} \frac{1}{2} \Delta_{\rho\mu} \Delta_{\rho'\lambda} \frac{\partial^2 E^{(0)}}{\partial a_{\rho\mu} \partial a_{\rho'\lambda}} + \cdots, \tag{B.26}$$

and

$$L_x = \sum_{\rho,\mu} \Delta_{\rho,\mu} \frac{\partial L_x^{(0)}}{\partial a_{\rho,\mu}} + \cdots . \tag{B.27}$$

The variation problem

$$\delta \langle \Phi | \hat{H}_B - \omega \hat{L}_x | \Phi \rangle = 0, \tag{B.28}$$

with $|\Phi\rangle$ being the coherent state, provides one with the linear equations in the infinitesimal $\Delta_{\rho,\mu}$ up to the leading order in the cranking frequency ω. Note that $a_{\rho\mu}^{(0)}$, $E^{(0)}$ and $L_x^{(0)}$ are the quantities in the stationary. Given that the energy expectation value in this dynamical case depends on $a_{\rho \pm 1}$ variables as $a_{\pi 1} a_{\nu 1}$, $a_{\pi 1}^2$ and $a_{\nu 1}^2$, the system is described by the following six linear equations:

$$\sum_{\rho=\pi,\nu; i=0,2} \Delta_{\pi 0} E_{\pi 0,\rho i}^{(0)} + \Delta_{\pi 2} E_{\pi 2,\rho i}^{(0)} + \Delta_{\nu 0} E_{\nu 0,\rho i}^{(0)} + \Delta_{\nu 2} E_{\nu 2,\rho i}^{(0)} = 0, \tag{B.29}$$

and

$$\Delta_{\pi 1} E_{\pi 1,\rho 1}^{(0)} + \Delta_{\nu 1} E_{\nu 1,\rho 1}^{(0)} = \omega L_{x,\rho 1}^{(0)}. \tag{B.30}$$

The terms $E_{\rho i,\rho' j}$ ($\rho \neq \rho'$) in Eq. (B.29) represent the second derivative of E with respect to $a_{\rho i}$ and $a_{\rho' j}$. We have only to consider the derivative at $\omega = 0$, where the energy has its minimum, the determinant of the coefficients $E_{\rho i,\rho' j}$ in the first four equations of Eq. (B.29) do not vanish. This leads to the solutions for $\Delta_{\rho,2}$ and $\Delta_{\rho 0}$ to the lowest order in ω as

$$\Delta_{\rho 2} = \Delta_{\rho 0} = 0 + O(\omega^2). \tag{B.31}$$

Appendix B: Formulas in the IBM-2 Framework

The $\Delta_{\rho 1}$ is obtained from Eq. (B.30):

$$\Delta_{\rho 1} = \frac{L^{(0)}_{x,\rho 1} E^{(0)}_{\rho' 1, \rho' 1} - E^{(0)}_{\pi 1, \nu 1} L^{(0)}_{x, \rho' 1}}{E^{(0)}_{\pi 1, \pi 1} E^{(0)}_{\nu 1, \nu 1} - E^{(0)}_{\pi 1, \nu 1}}. \tag{B.32}$$

By substituting Eqs. (B.31) and (B.32) into Eq. (B.27), the IBM moment of inertia (denoted by $\mathscr{I}_B$) is obtained:

$$\begin{aligned}
\mathscr{I}_B &= \frac{\sum_{\rho \neq \rho'} [L^{(0)}_{x,\rho 1}]^2 E^{(0)}_{\rho' 1, \bar{\rho} 1} - 2 E^{(0)}_{\pi 1, \nu 1} L^{(0)}_{x, \pi 1} L^{(0)}_{x, \nu 1}}{E^{(0)}_{\pi 1, \pi 1} E^{(0)}_{\nu 1, \nu 1} - [E^{(0)}_{\pi 1, \nu 1}]^2} \\
&= 2 \frac{\sum_{\rho \neq \rho'} A_\rho^2 \{(B_{\rho'} - A_{\rho'}^2)\alpha + C_{\rho' \rho}\}}{\sum_{\rho \neq \rho'} \{(B_\rho B_{\rho'} - A_\rho^2 A_{\rho'}^2)\alpha^2 + C_{\rho' \rho}(2 B_\rho \alpha + C_{\rho \rho'})\}} \\
&= \frac{(A_\pi^2 B_\nu + A_\nu^2 B_\pi - 2 A_\pi^2 A_\nu^2)\alpha + A_\pi^2 C_{\nu \pi} + A_\nu^2 C_{\pi \nu}}{(B_\pi B_\nu - A_\pi^2 A_\nu^2)\alpha^2 + (B_\pi C_{\nu \pi} + B_\nu C_{\pi \nu})\alpha + C_{\pi \nu} C_{\nu \pi}},
\end{aligned} \tag{B.33}$$

where $L_{x,\rho 1}$ presents the derivative of the expectation value $\langle L_{\rho x} \rangle$ in terms of $a_{\rho 1}$:

$$L^{(0)}_{x,\rho 1} \equiv \frac{\partial \langle \hat{L}_{\rho x} \rangle}{\partial a_{\rho 1}} = \frac{4\sqrt{2} N_\rho \beta_\rho}{1 + \beta_\rho^2} \sin\left(\gamma_\rho + \frac{\pi}{3}\right) \equiv A_\rho. \tag{B.34}$$

The term $E^{(0)}_{\rho 1, \rho 1}$ in Eq. (B.33) is given by

$$E^{(0)}_{\rho 1, \rho 1} \equiv \frac{\partial^2 \langle \hat{H}_B \rangle}{\partial a_{\rho 1}^2} = B_\rho \alpha + C_{\rho \bar{\rho}}, \tag{B.35}$$

with

$$B_\rho = \frac{64 N_\rho (N_\rho - 1) \beta_\rho^2}{(1 + \beta_\rho^2)^2} \sin^2\left(\gamma_\rho + \frac{\pi}{3}\right) \tag{B.36}$$

and

$$C_{\rho \rho'} = \frac{\partial^2}{\partial^2 a_{\rho 1}} \langle \varepsilon n_{d\rho} + \kappa \hat{Q}_\pi \cdot \hat{Q}_\nu \rangle, \tag{B.37}$$

where

$$\begin{aligned}
\frac{\partial^2}{\partial^2 a_{\rho 1}} &\langle \varepsilon n_{d\rho} + \kappa \hat{Q}_\pi \cdot \hat{Q}_\nu \rangle \\
&= \frac{4 N_\rho \varepsilon}{(1 + \beta_\rho^2)^2}
\end{aligned} \tag{B.38}$$

$$-\kappa\left[\frac{2N_\rho N_{\rho'}}{(1+\beta_\rho^2)^2(1+\beta_{\rho'}^2)}\left\{2a_{\rho',0}+\chi_{\rho'}(2a_{\rho'2}^2-a_{\rho'0}^2)\sqrt{\frac{2}{7}}\right\}\right.$$

$$\times\left(4a_{\rho 0}+\chi_\rho\sqrt{\frac{2}{7}}(1-a_{\rho 0}^2+6a_{\rho 2}^2)\right)$$

$$+\frac{8N_\rho N_{\rho'}}{(1+\beta_\rho^2)^2(1+\beta_{\rho'}^2)}a_{\rho'0}a_{\rho 2}\left(1+\chi_{\rho'}a_{\rho',0}\sqrt{\frac{2}{7}}\right)$$

$$\left.\times\left\{4+\chi_\rho\sqrt{\frac{1}{7}}(4\sqrt{2}a_{\rho 0}-(1+\beta_\rho^2)\sqrt{3})\right\}\right]$$

$$=\frac{4N_\rho\varepsilon}{(1+\beta_\rho^2)^2}$$

$$-\frac{2N_\rho N_{\rho'}\kappa}{(1+\beta_\rho^2)(1+\beta_{\rho'}^2)}\times\left[\left\{2\beta_{\rho'}\cos\gamma_{\rho'}-\chi_{\rho'}\beta_{\rho'}^2\cos 2\gamma_{\rho'}\sqrt{\frac{2}{7}}\right\}\right.$$

$$\times 4\beta_\rho\cos\gamma_\rho+\chi_\rho\sqrt{\frac{2}{7}}(1+\beta_\rho^2(4\sin^2\gamma_\rho-1))$$

$$+\sqrt{2}\beta_{\rho'}^2\sin 2\gamma_{\rho'}\left(1+\chi_{\rho'}\sqrt{\frac{2}{7}}\beta_{\rho'}\cos\gamma_{\rho'}\right)\left\{4+\chi_\rho\sqrt{\frac{1}{7}}(4\sqrt{2}\beta_\rho\cos\gamma_\rho\right.$$

$$\left.\left.\times-(1+\beta_\rho^2)\sqrt{3})\right\}\right].$$

The term $E_{\pi 1,\nu 1}^{(0)}$ on the RHS of Eq. (B.33) is given by

$$E_{\pi 1,\nu 1}^{(0)}\equiv\frac{\partial^2\langle\hat{H}_B\rangle}{\partial a_{\pi 1}\partial a_{\nu 1}} \tag{B.39}$$

$$=\alpha\frac{\partial^2\langle\hat{L}^2\rangle}{\partial a_{\pi 1}\partial a_{\nu 1}}$$

$$=\alpha\frac{64N_\pi N_\nu\beta_\pi\beta_\nu}{(1+\beta_\pi^2)(1+\beta_\nu^2)}\sin\left(\gamma_\pi+\frac{\pi}{3}\right)\sin\left(\gamma_\nu+\frac{\pi}{3}\right)$$

$$\equiv\alpha D.$$

Under the condition that the proton and the neutron have the identical values of the deformation parameters, $\beta_\pi=\beta_\nu=\beta$ and $\gamma_\pi=\gamma_\nu=\gamma$, one has

$$L_{x,\rho 1}=\frac{4\sqrt{2}N_\rho\beta}{1+\beta^2}\sin\left(\gamma+\frac{\pi}{3}\right), \tag{B.40}$$

Appendix B: Formulas in the IBM-2 Framework

$$E^{(0)}_{\rho 1,\rho 1} = \alpha \frac{64 N_\rho (N_\rho - 1)}{(1+\beta^2)^2} \beta^2 \sin^2\left(\gamma + \frac{\pi}{3}\right) + \frac{4 N_\rho \varepsilon}{(1+\beta^2)^2} \quad \text{(B.41)}$$

$$-\frac{2 N_\rho N_{\rho'} \kappa}{(1+\beta^2)^3} \times \left[\left\{2\beta \cos\gamma - \chi_{\rho'}\beta^2 \cos 2\gamma \sqrt{\frac{2}{7}}\right\}\right.$$

$$\times 4\beta \cos\gamma + \chi_\rho \sqrt{\frac{2}{7}}(1 + \beta^2(4\sin^2\gamma - 1))$$

$$+ \sqrt{2}\beta^2 \sin 2\gamma \left(1 + \chi_{\rho'}\sqrt{\frac{2}{7}}\beta \cos\gamma\right)$$

$$\left.\times \left\{4 + \chi_\rho \sqrt{\frac{1}{7}}(4\sqrt{2}\beta \cos\gamma - (1+\beta^2)\sqrt{3})\right\}\right]$$

and

$$E^{(0)}_{\pi 1,\nu 1} = \alpha \frac{64 N_\pi N_\nu}{(1+\beta^2)^2} \beta^2 \sin^2\left(\gamma + \frac{\pi}{3}\right). \quad \text{(B.42)}$$

Since we consider the axially symmetric case ($\gamma = 0°$), A_ρ, B_ρ and $C_{\rho\rho'}$ becomes simpler:

$$A_\rho = \frac{2\sqrt{6} N_\rho \beta}{1+\beta^2}, \quad \text{(B.43)}$$

$$B_\rho = \frac{48 N_\rho (N_\rho - 1)\beta^2}{(1+\beta^2)^2}, \quad \text{(B.44)}$$

$$C_{\rho\rho'} = \frac{2 N_\rho [2\varepsilon - N_{\rho'}\kappa(2\beta - \chi_{\rho'}\beta^2 \sqrt{\frac{2}{7}})\{4\beta + \chi_\rho \sqrt{\frac{2}{7}}(1-\beta^2)\}]}{(1+\beta^2)^3} \quad \text{(B.45)}$$

and

$$D = \frac{48 N_\pi N_\nu \beta^2}{(1+\beta^2)^2} = 2 A_\pi A_\nu. \quad \text{(B.46)}$$

By comparing the cranking moment of inertia of boson system $\mathscr{J}_B$ with the corresponding moment of inertia in the fermion system, the strength of the LL term α are then obtained by solving the quadratic equation with respect to α. This procedure generally gives two solutions, the physically-relevant one of which is taken.

B.2 Diagonalization of Boson Hamiltonian

M-Scheme

The IBM Hamiltonian can be diagonalized in a set of bases conserving M (z-component of the total angular momentum J) rather than in a coupled form with respect to J (J scheme). The merit of the M-scheme diagonalization is that it is much easier to compute the Hamiltonian matrix elements analytically than in J-scheme. Particularly when one calculates the overlap between the eigenvector and the corresponding intrinsic state, the M-scheme basis is quite feasible to handle. In a large-scale shell model diagonalization the dimension of Hamiltonian matrix inflates in the M scheme. Also in the IBM the M-scheme dimension is larger than the J-scheme to a certain extent, the computational cost is still moderate as compared to shell-model calculation.

The M-scheme basis for IBM considered in this thesis is given by

$$|\Phi_M\rangle = \frac{1}{\prod_{\rho,lm} \sqrt{N_{\rho,lm}!}} \sum_{\rho,lm} (b^\dagger_{\rho,lm})^{N_{\rho,lm}} |0\rangle. \tag{B.47}$$

Here $b_{\rho,lm}$ ($\rho = \pi, \nu$) represents the monopole ($l = 0$) and the quadrupole ($l = 2$) bosons with $m = 0$ and $m = \pm 2, \pm 1, 0$, respectively. $N_{\rho,lm}$ stands for the number of bosons with l and m. The denominator on the RHS of Eq. (B.47) is the normalization factor. The boson vacuum $|0\rangle$ is interpreted as an inert core. Thus the boson number is counted following the general rule in the IBM and is also a good quantum number.

For a set of the M-scheme bases, one calculates the matrix elements of the IBM Hamiltonian with a certain value of M with $|M| \leq J$. Note that the M-scheme bases having different values of M are not mixed, as they are orthogonal to each other. When constructing the Hamiltonian matrix elements, one has a constraint of $2(N_{\rho,22} - N_{\rho,2-2}) + (N_{\rho,21} - N_{\rho,2-1}) = M_\rho$ concerning the boson number, where M_ρ represents the z-component of the total proton and neutron angular momenta. Because we use the IBM-2, one has additional constraint of $M_\pi + M_\nu = M$. Also all eigenstates considered in this thesis have the positive parity.

The Hamiltonian matrix is decomposed into sub-matrices belonging to each value of M (cf. Eq. (B.48)).

$$H = \begin{pmatrix} \boxed{M=0} & 0 & 0 & 0 & 0 \\ 0 & \boxed{M=1} & 0 & 0 & 0 \\ 0 & 0 & \boxed{M=-1} & 0 & 0 \\ 0 & 0 & 0 & \boxed{M=2} & 0 \\ 0 & 0 & 0 & 0 & \ddots \end{pmatrix} \tag{B.48}$$

Appendix B: Formulas in the IBM-2 Framework

The eigenvector of the boson Hamiltonian is eventually written as a linear combination of M-scheme bases, and provides the excitation energies and other spectroscopic observables, which are identical to those obtained from the J-scheme diagonalization having good angular momentum and the particle number.

Calculation of Collective Wave Function for Boson System

Like the collective wave function in the Bohr Hamiltonian approach, the IBM wave function of interest can be considered in the $\beta\gamma$ plane, by taking the overlap between the eigen state and the corresponding coherent state. In the following, only the case either of proton or neutron is considered for the sake of simplicity, and thus index ρ is neglected. The coherent state is transformed as

$$|\Phi(\beta,\gamma)\rangle = \frac{1}{\sqrt{N!(1+\beta^2)^N}} (s^\dagger + \alpha_0 d_0^\dagger + \alpha_2 d_2^\dagger + \alpha_2 d_{-2}^\dagger)^N \qquad (B.49)$$

$$= \frac{1}{\sqrt{N!(1+\beta^2)^N}}$$

$$\times \sum_{N_s, N_{d_0}, N_{d_2}} {}_{N_s+N_{d_0}+N_{d_2}+N_{d_{-2}}}C_{N_s} \cdot {}_{N_{d_0}+N_{d_2}+N_{d_{-2}}}C_{N_{d_0}} \cdot {}_{N_{N_{d_2}}+N_{d_{-2}}}C_{N_{d_2}}$$

$$\times (\beta\cos\gamma)^{N_{d_0}} (\frac{1}{\sqrt{2}}\beta\sin\gamma)^{N-N_s-N_{d_0}} (s^\dagger)^{N_s} (d_0^\dagger)^{N_{d_0}} (d_2^\dagger)^{N_{d_2}} (d_{-2}^\dagger)^{N_{d_{-2}}} |0\rangle.$$

A set of the m-scheme basis $|\phi\rangle$ is given as

$$|\phi\rangle = \sum_i^{\dim} c_i |\phi_i\rangle, \qquad (B.50)$$

where i stands for the state index which runs from 1 through the dimension of the Hamiltonian matrix. Amplitude c_i is generated by the diagonalization of the Hamiltonian matrix. $|\phi_i\rangle$ is given as

$$|\phi_i\rangle = \frac{1}{\sqrt{N_s! N_{d_2}! N_{d_1}! N_{d_0}! N_{d_{-1}}! N_{d_{-2}}!}} \qquad (B.51)$$

$$\times (s^\dagger)^{N_s} (d_2^\dagger)^{N_{d_2}} (d_1^\dagger)^{N_{d_1}} (d_0^\dagger)^{N_{d_0}} (d_{-1}^\dagger)^{N_{d_{-1}}} (d_{-2}^\dagger)^{N_{d_{-2}}} |0\rangle.$$

One may then compute the overlap of the coherent state of Eq. (B.49) and M-scheme basis of Eq. (B.51). The non-vanishing parts in the overlap $\langle\Phi|\phi\rangle$ must fulfill

$$N_{d_{\pm 1}} = 0, \quad k = N_s, \quad l = N_{d_0}, \quad m = N_{d_2}, \quad N-k-l-m = N_{d_{-2}}. \qquad (B.52)$$

We then obtain the overlap

$$\langle \Phi(\beta,\gamma)|\phi\rangle = \sum_i^{\dim} c_i \langle \Phi(\beta,\gamma)|\phi_i\rangle, \quad (B.53)$$

where

$$\langle \Phi(\beta,\gamma)|\phi_i\rangle = \frac{1}{\sqrt{N!(1+\beta^2)^N}} {}_N C_{N_s} \cdot {}_{N-N_s} C_{N_{d_0}} \cdot {}_{N-N_s-N_{d_0}} C_{N_{d_2}} \cdot \delta_{N_{d_1},0} \cdot \delta_{N_{d_{-1}},0}$$

$$\times (\beta\cos\gamma)^{N_{d_0}} \left(\frac{1}{\sqrt{2}}\beta\sin\gamma\right)^{N-N_s-N_{d_0}}. \quad (B.54)$$

Taking into account the proton and the neutron degrees of freedom under the assumption $\beta_\pi = \beta_\nu = \beta$ and $\gamma_\pi = \gamma_\nu = \gamma$ (subscript B, indicating boson system, is omitted), the probability density distribution $\rho(\beta,\gamma)$ can be calculated by the equation

$$\rho(\beta,\gamma) = \mathcal{N}^{-1} \sum_i^{\dim} |\langle \Phi(\beta,\gamma)|\phi_i\rangle|^2 \beta^3 |\sin 3\gamma|, \quad (B.55)$$

where the normalization factor $\mathcal{N}$ is determined so that

$$\int_0^\infty \beta d\beta \int_0^{2\pi} d\gamma \rho(\beta,\gamma) = 1. \quad (B.56)$$

References

1. Greiner W, Maruhn JA (1996) Nuclear models. Springer, Berlin
2. Ring P, Schuck P (1980) The Nuclear many-body problem. Springer, Berlin
3. Bender M, Heenen P-H, Reinhard P-G (2003) Self-consistent mean-field models for nuclear structure. Rev Mod Phys 75:121–180
4. Bardeen J, Cooper LN, Schrieffer JR (1957) Microscopic theory of superconductivity. Phys Rev 106:162
5. Bardeen J, Cooper LN, Schrieffer JR (1957) Theory of superconductivity. Phys Rev 108:1175
6. Schaaser H, Brink DM (1984) Calculations away from SU(3) symmetry by cranking the interacting Boson model. Phys Lett B 143:269
7. Ginocchio JN, Kirson M (1980) An intrinsic state for the interacting Boson model and its relationship to the Bohr-Mottelson model. Nucl Phys A 350:31
8. Caprio MA, Iachello F (2004) Phase structure of the two-fluid proton-neutron system. Phys Rev Lett 93:242502
9. Caprio MA, Iachello F (2005) Phase structure of a two-fluid bosonic system. Ann Phys 318:454